I0467744

The Elements of Elven Magic:

A New View of Calling the Elementals Based Upon the Periodic Table of Elements

By The Silver Elves

Copyright © 2015 The Silver Elves, Michael J Love and Martha C. Love

The original photo (before graphic enhancements) on the cover is of a Japanese apothecary shop in Narita Town, Japan, and was taken in 2015 by Kiki Prapatat who has given us permission to use it for this book. We feel this traditional apothecary shop demonstrates substances of life—vegetable, animal, and mineral—that encompasses all of the elementals.

All rights reserved.

ISBN-13: 978-1519282736

ISBN-10: 1519282737

Printed in the United States of America by CreateSpace

Without limiting the rights under the copyright reserved above, no part of this publication may be reproduced, stored in or introduced into a retrieval system, or transmitted in any form or by any means (electronic, mechanical, by photocopying, recording or otherwise) without the prior written permission of the copyright owner and the publisher of the book.

DEDICATION

This book is dedicated to the sister of our blood, to the sister of our magic, to our faerie sister, Gail Avrill.

> THERE ARE MANY PATHS TO ELFIN AND EVEN MORE WHEN YOU ARRIVE.
> ...OLD ELVEN KNOWLEDGE

TABLE OF CONTENTS

A LETTER FROM A SISTER OF THE ELF QUEEN'S DAUGHTERS 15

INTRODUCTION 17

CHAPTER ONE: THE LORDS AND LADIES OF THE AIR AND THE NOBLE GASES, MASTERS OF THE NEBULOUS REALMS 21

Hydrogen, Lord Quaridarea 21

Helium, Lord Wevas 23

Nitrogen, Lady Moderea 25

Oxygen, Lady Emper 26

Fluorine, Lady Leporn 28

Neon, Lord Élgol 30

Chlorine, Lady Lyrrata 31

Argon, Lord Enil 33

Krypton, Lord Higas 35

Xenon, Lady Wyrys 37

Radon, Lord Alur 39

Ununoctium, Lord Murdorkere 41

CHAPTER TWO: THE LORD AND LADY OF THE LIQUID ELEMENT, THE MASTERS OF FLUIDITY 43

Bromine, Lady Stenad..43

Mercury, Lord Merku...45

CHAPTER THREE: THE LORD AND LADIES OF THE SOLID ELEMENTS, THE MASTERS OF ENDURANCE AND MANIFESTATION 49

Lithium, Lady Dongur 49

Beryllium, Lady Sokyncor 51

Boron, Lord Burnas 53

Carbon, Lord Benas 54

Sodium, Lord Soulcor) 56

Magnesium, Lady Brisylcor 58

Aluminum, Lady Nasnalney 59

Silicon, Lord Wiclor 61

Phosphorus, Lord Salfarsey 63

Sulfur, Lord Sulfur 64

Potassium, Lady Nårhemåcor 66

Calcium, Lady Cahyrcor 68

Scandium, Lord Nolcor 69

Titanium, Lord Jalcor 71

Vanadium, Lady Afrocor 72

Chromium, Lady Luharcor 74

Manganese, Lady Morten 76

Iron, Lord Feri 78

Cobalt, Lady Korsal 79

Nickel, Lady Maletil 81

Copper, Lord Dwartyn 82

Zinc, Lord Wiz 84

Gallium, Lady Romecor 86

Germanium, Lord Revancor 88

Arsenic, Lady Arnalas..89

Selenium, Lady Moncor 91

Rubidium, Lord Saråcor..93

Strontium, Lady Cordicor 95

Yttrium, Lord Vandcor 96

Zirconium, Lady Fadymcor 98

Niobium, Lady Unos 100

Molybdenum, Lady Setynmacor 101

Ruthenium, Lord Hawart 103

Rhodium, Lord Sascor 105

Palladium, Lady Hasarcor 106

Silver, Lady Arvyn 108

Cadmium, Lady Nacor 110

Indium, Lord Newor 112

Tin, Lord Naf 114

Antimony, Lady Matela 116

Tellurium, Lord Arthorcor 117

Iodine, Lady Rodfaso 119

Caesium, Lord Mercor 121

Barium, Lady Sylwima 123

Lanthanum, Lady Naena 125

Cerium, Lady Rongra 126

Praseodymium, Lord Genlascor 128

Neodymium, Lady Galåscor 130

Samarium, Lord Sylmel 132

Europium, Lord Nosor 133

Gadolinium, Lady Anomal 135

Terbium, Lord Zoncor 137

Dysprosium, Lady Diculto 138

Holmium, Lord Dutinu 140

Erbium, Lady Ansomel 142

Thulium, Lord Rodcor 143

Ytterbium, Lady Vancor...145

Lutetium, Lord Ariscor 147

Hafnium, Lord Dencor 149

Tantalum, Lord Harala..150

Tungsten, Lady Bramyl 152

Rhenium, Lady Radencor 154

Osmium, Lady Luwyn..155

Iridium, Lady Britonme 157

Platinum, Lady Latyna 159

Gold, Lord Urm 161

Thallium, Lord Poscor 163

Lead, Lord Emni 165

Bismuth, Lady Harwi...167

CHAPTER FOUR: THE LORDS AND LADIES OF FIRE, MASTERS OF THE REALMS OF RADIANCE, TRANSFORMATION AND ILLUMINATION.......... 171

Technetium, Lord Arficor 172

Promethium, Lord Pornacor 174

Polonium, Lady Bimocor 175

Astatine, Lady Unsato 177

Francium, Lady Untov 178

Radium, Lord Alucor 180

Actinium, Lord Metacor 181

Thorium, Lord Charcor 183

Protactinium, Lady Liurcor 185

Uranium, Lady Romicor 186

Neptunium, Lord Lolyni 188

Plutonium, Lady Radasit 190

Americium, Lord Zårduron 192

Curium, Lady Ancor 194

Berkelium, Lady Lonliyon 196

Californium, Lord Fluercor 197

Einsteinium, Lady Damducor 199

Fermium, Lord Famyn 201

Mendelevium, Lord Tosoto......................................203

Nobelium, Lady Lonicor 205

Lawrencium, Lady Mynodcor 207

Rutherfordium, Lord Nelbarcor 208

Dubnium, Lady Ruyntyn 210

Seaborgium, Lord Åtovåcor 212

Bohrium, Lady Danicor 213

Hassium, Lord Majicor 215

Meitnerium, Lady Nuficor 217

Darmstradtium, Lord Wihòcor 218

Roentgenium, Lord Oloråcor 220

Copernicium, Lady Telegålåcor 221

Ununtrium, Lady Murdorcor 223

Flerovium, Lady Soficor 225

Ununpentium, Lord Murdoråcor 227

Livermorium, Lord Ralicor 228

Ununseptium, Lady Eldåcor 230

CHAPTER 5: IN CONCLUSION 233

Will There Be Other Elementals? 233

Is It Necessary to Use the Magic Circle of Protection
and the Triangle of Manifestation? 233

ABOUT THE AUTHORS 235

> REASON SPEAKS OF WHAT IS.
> IMAGINATION WHISPERS WHAT COULD BE!
> —OLD ELVEN SAYING

> *LEARNING IS NOT A COLLECTION OF FACTS*
> *BUT AN ACCUMULATION OF EXPERIENCES.*
> —*ELFIN KNOWLEDGE*

A LETTER FROM A SISTER OF THE ELF QUEEN'S DAUGHTERS

THE STUDY OF THE SCIENCE OF ELEMENTALS IS DEEP, Silver Flame and Zardoa . . .

. . . and can occupy lifetimes. But, until you begin to understand these essential beings, life can be chaotic. The initiate needs to exert will and direction upon the elementals so they will do their job of smoothing the way and giving warning of dangers. None of this is done by exertion of the conscious mind. Rather the soul, sitting in the seat of true consciousness, dominates and guides all the aspects of life. The elementals hunger for direction. They ultimately crave a task—a sense of working in tandem.
Our I Ching was Oppression (#47) changing to Dispersion (#59). As always the spirits gave us a glimpse into their world. The I Ching spoke clearly to the idea of paying attention to the elementals if we want to move towards our goal of dangma and dharma.
Go with the flow of elemental life that dances all around you.

Te Amo,
Elanor Tooke {One of the founders of the Elf Queen's Daughters}

INTRODUCTION

It is a common practice in magic to speak of four, sometimes five essential elements, which are the basic formation of the beings or spirits who are known as the Elementals. In Western magic these four elements are viewed as Fire, Water, Air and Earth. In traditional Eastern magic, which is to say essentially Chinese, Korean and Japanese magic, basically Taoist magic, there are five elements conceived of as Fire, Water, Wood, Metal and Earth. These two systems are often brought together somewhat, but not entirely, in modern magical systems by adding an element of spirit or ether to the Western systems.

We elves view the nature of the world into elements as well, but we conceive of our elementals in coordination with the Periodic Table of Elements of which there are four basic states of being. These are Solid/Earth, Liquid/Water, Gaseous/Air and Plasma; we however, as we will explain later have used Radiant or Transformative/Fire for this fourth elemental state. The fifth element, the formulation of ether or spirit, is not in our view an element at all but rather the underlying energetic nature of all being. It is the Unifying energy that unites them and is thus connected to and is an expression of the Divine Magic.

Often the elements of fire, water, earth and air are also seen as the Elementals. Thus the air element is also the air elemental. (Sometimes these are viewed in the form of otherworldly spirits associating Fire with Salamanders [dragons might be better], Air with Sylphs, Water with Undines and Earth with Gnomes, although we think Trolls might be more appropriate. If one were to give an association for the element of spirit it would probably be Angels.) However, to these elves, the elementals are the individual formulations or spirits of the elements. Thus Hydrogen, Nitrogen and Oxygen are Elementals of the Gaseous/Air element. Gold, Silver and Platinum are elementals of the Solid/Earth element.

Of course, these elementals can at times change forms. Gases can become liquid, liquid solid or gas, solid liquid and so forth. These indicate relationships between the elementals. The degree of ease or difficulty in such a transition is an indication of the ease or complexity of that relationship. However, while the elementals do change states of being, they are listed and arranged here according to their basic form and most common state of being at normal room temperature and pressure. This is, in part, why we chose radiant or radioactive rather than plasma for the fire element, since while there are elementals that can reach the state of plasma there is no elemental that is plasmic at normal room temperature and pressure.

On the other hand, while we look at the elementals in this more scientific and, we think, accurate form, we elves appreciate the poetic aspect of the traditional formulations. Calling on the spirits of the Air seems to us to be somehow more elegant than calling on the spirits of Gas and thus we have no problem with the traditional formulation although we do conceive of them differently than they are usually pictured. We'd tend to call on the Spirits of the Nebulous realms rather than spirits of Air but calling them air spirits is not a problem for us. And calling on the spirits of water, earth and fire is also more poetic than calling on liquid, solid and radiant, although we very much like radiant.

Also, there is nothing wrong with calling to an element whether we call it Air or Gaseous or Nebulous and in that way calling to all the spirits, all the elementals, of that element. However, it seems to these elves that it may sometimes be more efficient to call to the particular spirits or elementals that affect a chosen situation or event. If you wish to increase your wealth you can call to all the Earth/Solid/Material spirits but it might be more advantageous to call to the elementals of Gold, Silver and Platinum, or even Copper. All things material are composed of the elementals. If you know the basic elemental composition of any particular thing you can gain greater influence over that thing. This is comparable to the idea that by knowing the True Name of something you can gain power over it, although we elves think and act more in terms of influence or enchantment than of power over or control. For us, magic is always about relationship and soulful connection and all things in the Universe are related either directly or indirectly through the Divine Magic that lives in potentiality in all being.

Some might point out that in modern times most currencies are not based upon Silver or Gold anymore and thus evoking wealth might involve more ephemeral forces and this is surely true. Not everything is directly composed of the elementals. What, for instance, would be the elementals to evoke to find love? This is the region of what is commonly seen in Western magic as the fifth element, the element of the ethereal, ephemeral and thus of spirit. Still, in evoking love we, as human or humanoid or other beings, tend to wish love to manifest in a physical form. So, too, in evoking wealth we call on the elementals that are valued in the world and represent wealth. Besides Gold, Silver, Platinum and so forth, one could evoke Carbon, which is the principle composition of Diamonds. Carbon thus rules diamonds.

Therefore, in evoking love one might evoke those elementals that compose a human or humanoid or elven being (assuming one wanted love in a humanoid form and not from their dog or cat) and adding to that, using the power of visualization, the ethereal or spiritual aspects that one desired in such a being such as kindness or intelligence that cannot be specifically relegated to elemental form.

One might ask why not simply deal with the Master element, that of the ethereal that is so closely involved with the imaginal/magical realms and surely, as we said, there is nothing wrong in doing so. However, most things that individuals wish to manifest in the world involve material things, just as love and wealth have material aspects. Most individuals, as we pointed out, wish love to manifest in a physical form, through a person or persons. Calling on the elementals associated with that form can be of great help.

Of course, there are things such as fame, which some might seek, that don't have a direct material corollary and yet one still wishes to be famous to people. It is unlikely that one wants to be famous among rabbits or horses. One wants to be known and admired by human/humanoid beings. While human beings vary in small ways, 99% of the mass of the human body is made up of six elements. These are oxygen, carbon, hydrogen, nitrogen, calcium, and phosphorus and thus these would be the elementals to evoke when dealing with anything involving humans.

So, too, if one wished to develop the power of levitation one might consider evoking helium, the element most noted for being lighter than air. As you can see, some imagination is involved here, but that is magic, isn't it? Therefore, we present for your consideration a new and elven way at pursuing the magical art of evocation of the Elements and the Elementals. May it serve you well and hopefully evoke your own imagination and stir your particular and individual style of magic.

CHAPTER ONE:

THE LORDS AND LADIES OF THE AIR AND THE NOBLE GASES, MASTERS OF THE NEBULOUS REALMS

Hydrogen, Lord Quaridarea
(pronounced Q – air – rye – dare – ree – ah)
Lord of the Invisible Fire
Chant: "Mighty Lord of Air and Fire brings us now our
heart's desire."

This magical spirit is the lightest of the elementals. This being so, this elemental may be of assistance to those seeking the power of levitation; however, as we pointed out in the introduction, Helium might be a better choice to evoke since the idea of being lighter than air is so commonly associated with that elemental and Hydrogen is explosive. Thus this elemental might push you to change more quickly than you may wish. Of course, there is no problem in evoking both these spirits if that is what you wish to do; however, if you are in a hurry and don't mind the stress then Lord Quaridarea is the one for you.

Lord Quaridarea also represents the most abundant chemical substance in the Universe so his reach is vast and wide. Therefore, if you wish to evoke the aspect of abundance in your life, this would be a good spirit with which to start. Since hydrogen is colorless, odorless, tasteless, this Lord is nearly invisible when summoned and requires a psychic sense to see him. One tends to feel his presence rather than see it. So if you wish to develop the power of invisibility, to go through the world unseen and yet have great influence in it, he is the elemental to consider. Further, he may also be of value in anything that has to do with ghosts or shards of the dead.

Hydrogen is a non-toxic, nonmetallic, highly combustible diatomic gas and thus while Lord Quaridarea might be thought to be mild mannered, he is really quite sensitive and volatile, manners and decorum are especially important in summoning him. He knows how influential he is. At the same time, Hydrogen is the simplest element, so put on no "airs" around this Lord, directness and simplicity will be appreciated.

And while he gets along well with water and many liquids and doesn't mind going down into the world, he doesn't get on well with metals and tends to make them brittle. Therefore, it is unwise to have anything metal in your circle when evoking him. If you feel you need a ritual blade or athame, let it be made of obsidian or flint or stone or even wood or bone. The same is true of your other instruments (tools) of magic.

On the other hand, for this same reason he is quite useful at weakening metal. In the olden days he may have been evoked to weaken the swords of one's enemies. And while we said that he might be useful in evoking abundance; it is unlikely that abundance will appear in the form of gold, silver or any other metal. However, if you wish things to flow to you, he surely can be of assistance there.

The word Hydrogen, which is from the Greek, means water forming. Since water is H_2O, two Hydrogen atoms combined to one Oxygen atom, having a chalice or bowl of water in your circle can be very conducive in his summoning.

Helium, Lord Wevas (pronounced Wee – vace)
Lord of the Spiritually Aspiring
Chant: "We call you Lord of mystic airs, floating free from
 worldly cares."

Lord Wevas rules Helium, one of the Noble Gases, that is to say a gas that tends to be inert and non-reactive chemically. In other words, he is inclined to be an introvert, feels quite satisfied in himself and pursues his own predilections with little care about any other spirits doings or opinions, however, since he rules over the second most abundant element, his influence should not be underestimated. He, like Lord Quaridarea, the Lord of Hydrogen, is nearly invisible. He is also the second lightest spirit in the Universe, yet because of his reputation has a greater association with levitation.

In liquid form, helium is used in cryogenics (the study of things at extremely low temperatures), especially in the cooling of superconducting magnets, with its main application being in MRI scanners. Even at absolute zero Helium doesn't become solid. Thus Lord Wevas is a very "cool guy" and has a very spiritual air. He is unlikely to aid you with most material concerns, with which he has little interest, but is very interested in spiritual and intellectual matters. He sees himself as a facilitator. He eases the way, makes magic happen with less friction and more efficiency. He is noted as a Master Planner. He sees to the depth of things and knows how to make things happen, to draw things to you. However, while he will ease the way, or inspire you as to how to get what you want with ease, he is unlike to get his own hands dirty. He has a somewhat low opinion of the material world in its solid states and thus he probably should not be evoked with any of the solid elementals. Nor is it necessary to have magical instruments in calling him. Visualization of one's tools and ritual is not only enough for him but he finds that much more impressive.

Helium is also used as a pressurizing and purge gas and as a protective atmosphere for arc welding. Thus Lord Wevas tends to instigate spiritual change in soulful individuals. He understands what you need in order to grow and develop and is not beyond creating pressure in your life to instigate growth in the right directions. However, know that if he does this he is also watching over you and will not let you fall as long as you persevere.

Helium comes from the Greek Helios, which means the Sun. Helium is used in processes such as growing crystals to make silicon wafers, which are used for super-conductivity and in solar cells. Therefore, evoke Lord Wevas in the full light of day rather than at night or even in the shade. This will make your evocation so much more powerful and make it much more likely that he will heed your call. However, he doesn't care for group evocations. It's not that you can't call him in a lodge or coven, but he's less inclined to respond. He prefers solitary witches and wizards, individual scientists, artists and hermits and of all of these he prefers eccentrics the best.

Unlike his brother Lord Quaridarea, Lord Wevas is not flammable. He is unlikely to explode in a temper and go off on you. Rather, the opposite is true, he is so absorbed in his own thoughts and researches that he is like an eccentric scientist or artist. Remember, he is an introvert and busy with his own work. You may need to call him a number of times to get his attention before he actually hears or heeds you.

Nitrogen, Lady Moderea
(pronounced moe – deer – ree - ah)
Lady of the Fields
Chant: "Enrich us all the best you can and make the earth
a better land."

Nitrogen is the seventh most abundant element in the Universe; therefore, Lady Moderea is very influential. At room temperature, it is colorless and odorless and thus like the previous two elementals, this Lady can be hard to see. It constitutes about 78% of the Earth atmosphere, so once again she is a very important being. And while Nitrogen composes only about 3% of the mass of the human body, it is none-the-less the fourth major element in the body after oxygen, carbon, and hydrogen. It is often a key ingredient in fertilizer, thus Lady Moderea is a patron of gardeners, farmers and others who grow things. So if you wish to provoke growth in your life or some situation in your life, she is a good elemental to summon.

Nitrogen is also a key part of ammonia, nitric acid and organic nitrates, which are used as propellants and explosives, and also in cyanides, so she has her dangerous side, too. She is, therefore, a key ingredient in fertilizer bombs. In this capacity, Nitrogen is perhaps best noted in the form of saltpeter, sodium nitrate or potassium nitrate (note those combinations of elementals), which is a primary ingredient in gunpowder. When compounds explode, burn or decay back into nitrogen gas, tremendous energy is released. The Lady thus must be treated with due care and respect. Interestingly, nitrogen also composes part of the compounds that make Kevlar fabric and super glue so she has her protective as well as bonding aspects as well. Find out the other elementals involved to summon a combination that will produce the aspects you desire.

Alchemists during the Middle Ages called nitric acid, a nitrogen compound, aqua fortis or strong water. Aqua regia or royal water, a mixture of mixture of nitric and hydrochloric acids, was noted for its ability to dissolve gold, which was important in alchemical work. Thus this elemental has a strong association with Alchemists.

Also, every major pharmacological drug class has Nitrogen molecules as a part of it. This includes antibiotics but also caffeine and morphine. You can see how very important this elemental is and the variety of things this Lady can influence is truly vast. She has influence over amino acids (and therefore proteins) and the nucleic acids, which is to say DNA and RNA. And despite its tendency to be used for explosive purposes it tends to extinguish fires. Its explosiveness, as we saw, coming from its tendency to wish to return to its elemental state. Thus, like Helium, Lady Moderea is really a bit of an introvert. She'd rather putter around a garden, minding her own business, and only tends to get volatile when forced into situations or relationships with which she's uncomfortable. Thus, we repeat that she loves to see things grow and is in her best mood when called in that capacity.

If you are someone who has been forced to be other than you truly are, pressured into conforming to standards that don't suit your true nature, than this elemental is the one to summon to help you return to your true s'elf and your true relationship with Nature.

෧෧෧

Oxygen, Lady Emper (pronounced eem - peer)
Lady of the Enthusiasm
Chant: "Awaken bright and stirring be, it's time for all to clearly see."

Unlike Helium and Nitrogen, whose elementals tend to be introverts, Lady Emper, the Mistress of Oxygen, is an out and out extrovert who interacts easily with all sorts of elementals. The name Oxygen, however, means acid forming, thus she can be a bit caustic and is not afraid of expressing her opinions or her wit. This Lady can party, but she can also burn you out. The expression burning the candle at both ends definitely applies to her. Ruling the third most abundant element in the Universe, she is extremely powerful and influential.

While Oxygen composes most of the mass of living organisms, this elemental also has great influence over many inorganic compounds, such as teeth, shells and bone. But, Oxygen can be toxic to some forms of life particularly those that were dominant early on in Earth's life cycle. Over twenty percent of the volume of air is composed of Diatomic oxygen gas. Without the influence of this elemental, we could not live in human/humanoid bodies. Oxygen is also the most abundant element in the Earth's crust, so you can see how very important this elemental is.

Like its brothers and sister before it, it tends to be colorless and odorless at room temperature. Thus we have another spirit that is difficult to see directly. In air, it tends to be O_2 which is to say that really there is not one elemental but two sisters who are twins and can be easily mistaken for each other since they are essentially identical. So keep this in mind when summoning this elemental, you may get one or the other, but most likely both of them. They go by the same name. They love to hang out together. If you wish to deal with just one of them, have a bowl of water that is exposed during the evocation to direct sunlight. Otherwise, if you wish both sisters, you may call them at night. Be lively, they love energetic people, they love parties, they are 24 hr. party people. These elementals are enchantresses and tend to be hostesses. If you wish to get a coven or group together, they can be of great help.

If you can get them to materialize either in solid or a liquid form, they will probably appear to be sky blue in color. Their manifestation may be the inspiration for many Hindu gods and even elves being pictured as having blue skin.

Fluorine, Lady Leporn (pronounced lee - porn)
Lady of Affinity
Chant: "Together let us ever be, so none may ever sunder we."

Fluorine is the lightest of the Halogens. Halogens have the capacity to take on three or four different states at standard room temperature and pressure, thus the Lady Leporn is a shape shifter and can aid any, including actors or those who seek to be masters of disguise, who call on her to increase their ability in that skill. She is also the master of camouflage. Unlike her predecessors, she can be easily seen, if she so desires, usually appearing dressed in yellow.

While Fluorine is 24th in abundance in the Universe, it is 13th upon the Earth, so Lady Leporn's power is greater concerning material and worldly desires than most cosmic ones. She is an extreme extrovert and loves to interact with others. She even hangs out with the Noble gases who tend to be snobs toward nearly everyone else. However, when in the company of Hydrogen she becomes acidic. She is deeply affected by her companions. She is extremely attractive. Thus she might be seen as the patron of femme fatales. She is a bit of a *Bond girl* you might say and she is very adaptable. The *Fluo* in her name comes from Latin and means flow, for she tends to put people at the ease and help them be more flowing and flexible. This is particularly true of many the metal elementals that can be such hard asses otherwise. She loosens them up, so you might consider that when summoning some of them.

However, Lady Leporn is also extremely loyal to those with whom she bonds. Any attempt to separate her from her friends will meet with strong and potentially deadly response. Therefore, if you summon her you may very well find that she arrives with a friend or two or an entire crowd or perhaps party would be a better word. And again, you will know her by her love of the color yellow, sometimes appearing as pale yellow or yellow green. However, in her gaseous and individual state, she can be highly toxic, so if she shows up alone, be sure you are within your magic circle, since she doesn't like to be alone and can be in a very bad mood in that situation. She is much more pleasant when she is with others.

She is also somewhat of a high-maintenance individual. Her aid may cost you quite a bit of energy and effort. Before you make a compact with her, find out what it will cost you. On the other hand, if you do reach an agreement, know that she will be utterly faithful to it no matter what happens and will see things through to completion. This is a Lady that can take the heat and will resist any attempts, thus counter-spells, to make her deviate from her course.

She is, perhaps, best known for counteracting tooth decay. This lady has a great smile and a wonderful set of teeth.

When she comes in contact with Carbon she tends to heat things up and is inclined to add to the *greenhouse gases* that are such a current problem on the Earth. However, if you wished to heat up a colder planet, such as Mars, these two could be of great help. They can also heat up situations in your life, if that is your desire.

If you use crystals in your magic workings, you should know that Lady Leporn loves Fluorite, Fluorapatite, Cryolite, and especially Topaz. She will most likely be wearing Topaz when she appears, which is another way of distinguishing her from those in her company.

ॐॐॐ

Neon, Lord Èlgol (pronounced L - goal)
Lord of Omens
Chant: "Show us the way, oh, Lord of Light, guide us
through the darkened night."

Neon is another of the Noble gases, those elementals that tend toward introversion and are uncomfortable interacting with others and particularly are not fond of crowds unless it is of their own kind where they can all hang out together and never communicate, rather like the members of the Diogenes Club in the Sherlock Holmes stories who would gather together and never speak a word to each other.

Under standard conditions, Neon is another colorless and odorless gas, thus it tends toward the invisible. However, if Lord Èlgol does appear it will most likely be with a sort of red aura. You may not see him but will see the reddish hue around him or everything will turn a bit reddish upon his appearance. And while Lord Èlgol rules the fifth most abundant element in the Universe, his appearance on Earth is quite rare. He is a bit hard to get ahold of and likes to deal with spiritual matters and cares little for earthly concerns. Calling him to help with your worldly aspirations is truly a waste of your time and he will surely see it as a waste of his. And he can be highly volatile, so tread carefully.

Neon comes from the Greek word for *new* and Lord Èlgol's reputation for leading one to new things is legendary. He is the lord of signs and omens. If you need a sign about your spiritual path, than he is the one to call. This is his specialty and while he may not appear directly, he will respond with a sign. If it comes from someone wearing red or reddish orange, or who has red hair, or in some place or situation that have those colors it may be seen as particularly powerful and favorable.

He is also the patron of prophets, fortune-tellers, soothsayers and those who do augury. Therefore, if you wish to learn the Tarot, I Ching or any other form of divination, he's the elemental to call.

And while Lord Élgol is an introvert, he is highly susceptible to getting excited. He likes to be entertained so anything you can do to make your summoning electric and exciting will be very much appreciated by him and, of course, much more likely to attract his attention. A dull, serious, ponderous ritual is more likely to repel than attract him. The shamanic formula of ecstatic awakening, chanting, drumming, dancing etc. is a good way to go. He is very fond of *Voodoo* type rituals. In a traditional setting, he'd much prefer church services where individuals sing and wave their arms around, or even babble in *tongues*, than one where everyone sits or stands primly in pews bored to tears.

However, Lord Élgol's aid can be a bit pricey. So be generous with your energy when calling him. Give him a lot of song and dance, make a *play* for him (we mean this both ways), make him feel he is worth the effort and he will give you the guidance you need. He needs to be romanced in a sense, he needs to be charmed and valued, made to feel worthwhile. He knows how rare he is here.

ళళళ

Chlorine, Lady Lyrrata (pronounced ler – ray - tah)
Lady of Purification
Chant: "Clear the air and make us safe, protect us from the evil wraiths."

Lady Lyrrata is another shape-shifter, although some might say she is just very moody or even suffering from OCD (obsessive compulsive disorder) prone at times to get into a cleaning frenzy. When summoned she will

31

tend to appear with a yellowish green aura. The name Chlorine comes from the Greek for *pale green*. Also, she tends to have that very distinctive bleach-like smell, so you will know her when she comes and not mistake her for her cousin Fluorine's Lady Leporn. Having her arrive is a bit like the smell at a public swimming pool. Also, since essentially no chlorine was created in the Big Bang but rather was created from supernovae, Lady Lyrrata is far younger than Hydrogen, Helium, Oxygen and numerous other gaseous spirits.

She is principally noted for her purification, cleansing and protective abilities. Sodium chloride, commonly know as salt, has long been used in magic for the ability to clear crystals of previous vibrations and to protect a magic circle or home. Placing salt along doorways and windows is a common practice for those who wish to ward off negative forces and malefic spirits.

Chlorine is the 21st most abundant element found in the Earth's crust, so while this is not the most influential spirit, she is not without power and influence. It is also noted for its bleaching and disinfectant uses, but again that reinforces the idea that this spirit helps us purify, cleanse and protect ours'elves and our realms.

Chlorine ions are necessary to all known species and life forms, however, organic molecules that contain chlorine such as chlorofluorocarbons are part of the cause of ozone depletion; therefore, we need to understand that one can over cleanse, over protect, so that it is not only wicked spirits that are kept from your realm but all spirits whatsoever. This would be like having a church where only the holy, the pure and the perfect are allowed to enter. It would be a pretty empty church and would counteract the transformational purpose of spiritual devotion. Moderation and balance are important in magic as well as in life in general and we need to remember magic is in many ways about transforming situations from those that are less conducive to our visions and goals to those that foster our dreams and desires.

At high concentrations, elemental chlorine is toxic and deadly for all living organisms. In the first World War, it was used as a gaseous chemical warfare agent, so terrible and unpredictable, since the winds tended to blow it back upon those who released it, that both sides agreed to ban its use in warfare. So remember that while Lady Lyrrata can be very helpful, part of her nature makes her dangerous not only to those we wish to keep away from our magic circle but, if over evoked, to ours'elves. Call on her when needed, but balance her help with positive evocations as well. This is to say that while it is important to purify and protect one's circle, we should not forget the outward and visionary aspects of our magic; the goals, spiritual and material, we desire to achieve. If one did nothing but evoke protective spirits one would probably end up with agoraphobia and never leave one's home or circle.

Argon, Lord Enil (pronounced E - nile)

Lord of Stability

Chant: "Settled down and stable be my life upon the
swirling sea."

Argon, another of the introverted Noble gases, is the third most prevalent gas in the Earth's atmosphere so Lord Enil is a pretty influential guy. Almost all argon comes from the Earth's crust through the decay of potassium-40. In other words, Lord Enil has experienced a long transformational process and has come to a more perfected and stable state and he feels that very strongly. So if you need stability in your life, he is a valuable spirit to summon. However, note that his stability is not that of a rock that tends to resist change, but a gas. While Argon is a Noble gas and feels rather complete unto itself, it is still a gas, that is to say it is very adaptable in terms of its shape. What Lord Enil promotes is the

33

ability to move through chaotic situations, where there are many pressures and demands placed upon one, without getting deterred from one's course or upsetting one's essential feeling of balance and well being. This spirit and the others of the gaseous realms that follow are all very comfortable with many of the luminous/transformational/radioactive spirits, since those spirits may be seen as their ancestors or perhaps more accurately as those spirits that are now going through what they had been through in previous incarnations. The expression, *you remind me of mys'elf when I was young*, may most aptly describe their relationships with these spirits.

The name Argon comes from the Greek word meaning lazy or inactive, however, we think it might be more accurate to describe Lord Enil as self satisfied and confident. He is a *what you see is what you get* sort of guy and is unlikely to change his attitude or response for anyone. If you wish to feel more confident in yours'elf and your path, then he can be of great help with this aspiration. If you wish to make a vow concerning your magical vision and aspiration, he will help you endure to achieve it.

Lord Enil stabilizes situations and those around him. His presence has a great calming influence on other elementals who would react more dramatically under pressure if he were not around. Therefore, in calling on some elementals, he is a good one to call first, for he will help protect your circle, particularly when you are calling dangerous spirits or expect the circumstances to become potentially volatile. He settles things down. Stands as a protective shield between you and the forces that might seek to harm you. He is generally unperturbed by outside influence.

Since Argon is colorless and odorless, he is usually invisible; however, if summoned, he may appear as a blue-green gaseous light. He gets along well and easily with water, which is to say the threesome of one of the Oxygen sisters and two Lords of Hydrogen. Argon is also nonflammable and nontoxic as a solid, liquid, and gas, so Lord Enil is unlikely to be a danger to you. He's been through his turbulent phases of evolution and has become a very dependable and trustworthy spirit.

Krypton, Lord Higas (pronounced high - gace)
Lord of the Occult
Chant: "Reveal to me the secret ways endless till the end
of days."

Krypton comes from a Greek word that means *the hidden one*. Lord Higas, another of the noble gases, those evolved elementals, is therefore the lord of all things hidden and of the occult, not just in terms of esoteric knowledge or secrets of magick, although he surely rules and reveals those, but all hidden things: hidden treasure and other hidden things and secrets including secret societies of every sort. He is also the lord of what is rare and precious. He rules a colorless, odorless, tasteless gas and is thus usually invisible, which suits his nature well. However, if he does appear it may be wearing or emitting many colors, although the strongest or most dominant will likely be yellow and green, especially if he happens to come in the company of others, although that is unlikely to occur if you don't summon them together. If he arrives by himself he'll most probably manifest as a bright flash of white light, like a flash bulb from a camera, that leaves you momentarily blinded.

Like the other noble gases, Krypton has found use in lighting and photography, therefore Lord Higas helps illuminate situations and preserve or renew memories. He and the other noble gases by their natures seek to enlighten and uplift.

Of course, Krypton is most popularly noted for its association to Kryptonite that substance that renders Superman powerless. While this may seem an idle or frivolous connection,

35

it is not. Lord Higas reminds us that everyone, no matter how high or powerful, has a weakness; everyone is evolving toward greater being. No one can be all-powerful without being united with the whole of the Divine Magic, which seeks fulfillment for us all for it is part of each and everyone.

Krypton is one of the products of uranium fission, thus, like the Lord of Argon, Lord Higas is a transformed being who has been through a lot to reach this more stable and perfected state. He is a spiritually evolved being but prefers to act in secret rather than emphasize that fact publicly.

In solid form, Krypton takes on a crystalline structure. Crystals are the radios of the Universe. They send forth energy waves of information and vibration, thus he is related to the messenger gods and the gods of magic: Mercury, Hermes and Thoth and Isis and also the Malachim, the angelic messengers and emissaries. Therefore remember that Lord Higas has his own goals and destiny and helps the magician in order to fulfill his own mission, which is to enlighten. If you are one who helps uplift and enlighten others then you will surely receive his aid.

Unlike Lord Enil, however, Lord Higas is not fond of water and can become volatile when encountering this compound. So have no water in your circle when summoning this elemental or things could get dicey. If he has to, he will form a relationship with Fluorine's Lady Leporn, but he's not all that fond of her either. He'd rather work alone and he'd much rather be in Space, working on a Cosmic scale, than doing anything related to the Earth, so spiritual and cosmic aspirations only when seeking his aid. While he knows how to find secret treasure, he is unlikely to tell you how to do so unless it somehow promotes your spiritual evolution and destiny.

And be in your magic circle, even if only visualized, when summoning him, while he exudes no or little toxicity, he has a narcotic and hypnotic affect on the individual when in direct contact. He can cause the conjuror to fall asleep and possibly fall into a coma, rather like a heroin overdose or carbon monoxide asphyxiation, for he is eager to take you to another life.

ॐॐॐ

Xenon, Lady Wyrys (pronounced were - riss)
Lady of Cosmic History
Chant: "Tell us of the ages past, and how life formed and came to last."

Lady Wyrys is the historian of the Cosmos. She knows everything that needs or can be known about its formation and development. However, since Xenon is a colorless, odorless and dense gas, she not only tends to be invisible and very far removed from material concerns, but can also be difficult to understand. Her density doesn't mean she is dense in the sense that people use it to mean stupid, but rather she is incredibly complex in her explanations. There is a lot to the Cosmos.

She is also a patron (matron?) of space travel and thus of astronauts. She is additionally a promoter of Science Fiction as it encourages individuals to think about the future and future possibilities as well as our greater relationship with the cosmos. She further has transformational powers and can increase your own magical abilities making them more potent and effective.

The name Xenon comes from a Greek word for stranger, foreigner, or guest, and thus she has power over extraterrestrials but that can be extended to mean any person who comes from afar as a guest and visitor. Therefore, she can also be called upon when you travel to foreign lands, particularly to places you have never been previously and are thus newly exploring. She will see that you are protected and treated well, as long as you, as a host, treat others in such a situation with courtesy and care. Note the relationship to the

37

word Xenophobia, which is the fear or dislike of that which is foreign or strange. She helps one overcome xenophobia in ones'elf and in others.

She has a relaxing influence on those she encounters and is able to relieve one of one's care, pains and woes, therefore one can go through many otherwise tense situations with aplomb. She helps ease the stress of travel and change. If you have migrated to a new area, she will aid you to fit in. If you suffer from jet lag, she will help you revive and reenergize.

If she appears to you, you will probably notice a change in pressure, a feeling that things have just gotten heavier. If you see her it may be as a bright flashing or strobe-like light. She has a positive relationship with metal, so you may have metal in your circle and if you have a good deal of it she is likely to appear as being blue or lavender.

She has a fondness for mineral springs, so if you have some mineral water in your circle or in the triangle of manifestation, that will help her to appear. However, remember she is a cosmic being, even when she appears in material form, and she may charge you a bit for her aid. This cost will be in terms of spiritual effort and evolution, so meditation, yoga and other spiritual practices may be dedicated to her in order to energize her. She also has an affiliation with the planet Jupiter and all things that planet represents. (We recommend Stephen Skinner's *The Complete Magician's Tables* for magical associations and correspondences.) You may wish to have the symbol of Jupiter in your circle and/or the associative incense and other tools beloved of that planetary spirit. And if you make a request of her, the response will most likely come through a stranger.

She also has a pleasant relationship with Lady Emper, the ruler of Oxygen, and also has a fondness for quartz, so having quartz crystals in your circle can be attractive to her. She further has a working relationship with Lady Leporn, the ruler of Fluorine, although if evoking these two it is better to do so in the day under direct sun or at night or indoors under ultra-violet light.

Radon, Lord Alur (pronounced a - lure)
Lord of the Dark Arts
Chant: "Speak to us of secret lore, so we may shift
   our inner core."

Radon is a gas that is created from the decay of Radium, which itself is a decay product of Thorium and Uranium that will themselves eventually evolve into lead (an interesting note for those interested in Alchemy). However, Radon is still radioactive and thus still in transition from the luminous elementals to the elementals of gas and therefore has a close association with the elementals of fire/transformation and could very well have been placed among the spirits in that elemental state. Like many of its kindred gas/air spirits, it is nearly invisible. Lord Alur, the Master of Radon, is also the master of the Dark Arts. This is not to say that he does Dark Magic, although he certainly has in the past, but that he knows nearly everything that can and needs to be known about these arts. He is a reformed, or we might more accurately say reforming, spirit.

When one comes to understand the true nature of the Universe and how it functions one realizes how unwise it is to use the Dark Arts for personal gain, revenge, etc. However, even the Lords of Light need to understand the Dark Arts and their ways, in order to counteract them. Thus, Lord Alur will teach you everything you need for magical and psychic self-defense, how the Dark forces work, how to avoid their many lures and traps, and in what ways and situations one can use these arts if necessary without evoking severe karmic payback.

He also has power and knowledge concerning everything of the underworld in all its meanings from the world of the Dead of Hades to the workings of the Mafia and other criminal organizations to the underworld of Faerie. He has an association with the Goddess Hecate. He further knows everything that needs to be known of the Unseelie Fae.

39

This elemental is a dangerous being. He is rather like Aragorn in the form of Strider when the Hobbits first met him. He will take you through the darkest and most dangerous of territories under his protection while teaching you to protect yours'elf.

Radon is a very dense gas and unlike Lady Wyrys for whom this represents a very complex world and cosmological view, Lord Alur is still very attached to the goings on of the Earth and evolutionary progress there. He is sometimes caught between the urge to uphold and the temptation to violate the *Prime Directive* from *Star Trek* (thou shall not meddle with the internal workings of alien civilizations). He can't help wishing to influence primitive civilizations, like those on Earth, rather than allow them to evolve at their own pace and potentially destroy themselves.

You definitely wish to be in your magic circle for this evocation and using a magic triangle would be wise as well. Direct contact with this elemental could lead to lung cancer. And as we said, he is nearly invisible and very hard to detect. Still, if you summon him in a cool or cold environment he is likely to appear as yellow or orange yellow, depending on how cold it is. The colder it is the more orange he will appear. You may wish to call him outdoors in Winter. At the same time, he has a fondness for hot springs, so putting a bowl of hot mineral water out to evaporate in the cold air can be very appealing to him.

He is also another noble gas and thus doesn't much care for being evoked with other spirits. He's a loner to be sure. Some might think him slow or ponderous, since he is a very dead serious being, but he is just slow to anger and not easily aroused by the excitability of others. Although, he is more likely to react than Xenon's Lady Wyrys, who is a very distant being.

Alas, he is not fond of dwarves or any who mine the Earth. He'd rather it stay in its natural state than be torn up. Therefore, he is a patron to those who oppose fracking and those who protest the destruction of mountains to obtain its raw minerals. He actively counteracts the Dark Arts with his own deadly force. He is thus a patron of Kobolds, especially those that pester miners.

Remember, this is a very dangerous spirit. But if you are under attack from the dark forces, he would be more than happy to help sort them out for you.

ふふふ

Ununoctium, Lord Murdorkere
(pronounced muir – door – keyr - ree)
Lord of the Unknown
Chant: "I feel struck and need to change, my life I seek to rearrange."

Ununoctium is a highly unstable, synthetic element, which means it was not created in the Big Bang or supernovae but by humanity. In fact, some scientists are not even sure this element actually exists as yet, thus even the name Unumoctium is merely temporary. In a certain sense, this is a truly imaginal being. But while scientists are not yet certain, we magicians have delved into the dark and know that beings are there, waiting in potential. This element is still evolving from its luminous, radioactive state of being and has not even decided whether it will stabilize as a gas or a solid, so Lord Murdorkere has strong associations with all three of these elemental states.

If you are uncertain of what you are doing, what you wish to do with your life, what is coming next and whether you should go forth and attempt to influence the world or to retreat and continue to develop yours'elf in private, then you will surely find a sympathetic ear in Lord Murdorkere who is going through the same indecision.

However, if you feel struck in life and think there is no way out of whatever situation you find yours'elf within, then

41

Lord Murdorkere can be of help, although evoking him may have uncertain consequences. Your life will surely change, of that you can be assured, but the manner of that change and the amount of chaos involved is unknown. He will get you where you wish to go eventually, but it may be a long, strange journey to get there. In evoking him, you are conjuring the Unknown.

Lord Murdorkere is living proof that the Cosmos is still evolving and there are still many things that have yet to be discovered or created, therefore, he is the patron of all those who experiment, whether in science or art. However, he is a very rare being, a true eccentric, and is fond of those who dare to explore and challenge the unknown as he does.

Naturally, be in your magic circle when calling him, with him in a magic triangle. If he does appear he will come and go rapidly. There won't even be time to speak to him. Merely, have what you desire written on a piece of paper, or have a visualization board, or a talisman dedicated to your goal for when he comes it will be like the Flash zipping through. He'll see your talisman or whatever, energize it and be gone again almost before you realize he's arrived. Most likely, you won't see a thing. You'll merely feel that someone or something has passed by very rapidly. Still, when changes begin to occur, and this will also most likely happen relatively quickly, hold on, you are in for the rollercoaster ride of you life.

CHAPTER TWO:

THE LORD AND LADY OF THE LIQUID ELEMENT, THE MASTERS OF FLUIDITY

There are in fact only two Liquid elementals and Water, which is a compound formed, as you know, from two gaseous elementals, isn't one of them. However, since the compound Water is born of a marriage, the idea expressed in traditional magic that *Water* is associated with relationship makes sense. At standard temperature and pressure only Bromine and Mercury take liquid form, although the elements Caesium, Rubidium, Francium and Gallium become liquid at or just above room temperature so these are close associates of this elemental form and can be summoned under this element under certain circumstances and conditions.

Bromine, Lady Stenad (pronounced stee - naid)
Lady of the Unwanted
Chant: "I would to clear my karmic debt, upon a new
　　　course I am set."

Lady Stenad is the protector of the little people in all their forms, the poor, the homeless, the disadvantaged and every being who has found that life hasn't been quite fair to them. She is also a matron of Brownies, thus she appears in reddish brown. Her name comes from the Greek meaning strong smelling or stench, thus in part her association with the homeless. However, the saying *you really clean up well* can be applied to her. She acts to clear away and transform the negative into positive elements.

She is very much associated with the water spirits, but not so much with mermaids as with nixies and other dangerous water creatures, such as sharks and barracudas and even with Pirates. Therefore, she also has association with the unconscious aspects of being, particularly those aspects Jung would call shadow elements, those things we so despise about ours'elves that we don't even acknowledge that they exist in us and instead we project them upon others in judgmental forms. Thus she also deals with our past karma, the karma of our ancestors and all those things we did in the past that we regret and are not entirely proud of having done and are so eager to forget or have others forget. She will help you clear your karma and transform regret into the wisdom of experience. If your past has caught up with you, she can be of great help.

She has a sisterly association with Chlorine and Iodine. She is kindred to the compound salt. And while she can be corrosive and toxic, particularly when she is alone, she doesn't care to be by herself and almost never is. She loves the sea and thus salt water. So if you conjure her, have some seawater, or if not that, salt water in the triangle of manifestation. Note, however, she has sedative qualities, she can knock you out for a long time, and she also has anti-convulsive powers, so at least you will stop shaking if she happens to terrify you. If you are in a situation that has made you afraid, she will calm you down.

However, she evaporates quickly at room temperature, so don't expect her to stay long. Be brief, state your desire and get the ritual over. No long ceremony required here. If you wish to pay her, which is very wise, do so by giving generously to the poor or homeless. If, however, you are someone who looks down on the homeless and has no compassion for their situation, then she is unlikely to have any sympathy for your problems.

Under a bit of pressure Bromine becomes a metal, so don't try to pressure her, or she may become a hard ass. At the same time, if you bring her together with any metals and with water at the same time, things will start happening very quickly. Messages will fly, directives sent forth and so on. So if you wish to have things happen rapidly, have a piece of metal sitting partially in a bowl of water when you summon her. However, she has an antagonistic relationship with aluminum, so use caution there.

She was one of the earliest promoters of photography and helped with the development of the daguerreotype. Therefore, photos and vision boards are a good way to communicate your desires to her.

ॐॐॐ

Mercury, Lord Merku (pronounced mere - coo)
Lord of the Words of Magic
Chant: "Words of power I would know, my visions all the
world to show."

Lord Merku is the master of Words of Power, also of sigils, cyphers, glyphs, secret codes, runes and magical scripts and languages of all sorts from Tolkien's Sindarin and Quenya, to our own Arvyndase (see *Arvyndase (Silverspeech): a short course in the magical language of the Silver Elves*), and Dee's Enochian. Lord Merku is an enchanter and everything to do with influence through communication is his province, including public speaking, sign language and those who translate from one language to another.

Mercury, which is also commonly called Quicksilver (a nice elven name that), is the only metal that is liquid, although Caesium, Gallium, and Rubidium are metals that become liquid at just a little above room temperature, so there is an association there. In warmer climates they may be evoked as liquid elementals.

Mercury is, of course, poison to the touch, also toxic in gaseous form and when found in seafood. The saying *mad as a hatter* comes from the fact that those who made hats in the past sometimes used Mercury in doing so and slowly went crazy. Therefore, know that Lord Merku can talk your ears off until you feel like you are going nuts. Be direct, clear and succinct in communicating with him or you will get caught up in a conversation that may not easily end. You will start hearing voices, or one voice really, and people will think you have gone insane.

Unlike most other metals, Mercury is a poor heat conductor, however, he is an okay transmitter of electricity thus Lord Merku can take the heat and is reliable in passing on messages. Despite his fluid nature, he is a pretty stable and dependable fellow as well as being quite unique in his way. He is fond of the clever, the witty and the eccentric, particularly if they are brilliant eccentrics. He likes mad scientists and genius artists.

He is beloved of alchemists who called Mercury the First Matter and thought all other metals were formed from it, but this may be in part because Mercury is very similar to Gold, which is just left of Mercury on the Periodic table and whose composition is quite similar only with one less electron. They are kindred spirits and you will note that the Star Trek idea of the reformulation of atoms to create anything one desires is just another form of alchemy. Some day we will easily strip or add electrons to transform one element to another and then we will truly be Master Alchemists.

And while Lord Merku is fond of wit, he doesn't much care for sarcasm, caustic wit, cynicism or cutting remarks, especially if they are used to strain relationship between individuals or cause a disruption in communications.

He gets on well with most other metal elementals, except for Iron, who he doesn't much care for. This may well be, in

46

part, the reason that elves, so often associated with Silver or quicksilver, are said in Faerie Lore to be harmed by Iron in the same way that Kryptonite is said to weaken Superman. Although, Lord Merku isn't harmed by Iron so much as bound by Iron. He doesn't like to be limited, being the free flowing spirit he is. However, if for some reason you need to restrain him or threaten him for non-responsiveness (which is very unlikely) Iron will do the trick. So you may wish to have an athame or sword or wand of Iron in your circle but keep it covered up unless you need to use it. Also, Platinum, especially, and a few other metals are a bit snooty toward Lord Merku and think they are too good for him. So don't call them together unless you specifically need to do so.

Mercury, on the other hand, has a corrosive influence on Aluminum, so Mercury is not allowed aboard aircraft for fear of their association. He does not fly; he flows.

When summoning Lord Merku you could have some cinnabar, which is a red, quartz like solid, composed of mercury sulfide, which is the source of the color vermillion, in the triangle. However, know that it can be toxic, so caution and proper handling is required. However, you can find it in the form of jewelry that has been lacquered to prevent toxicity. Alternately, you might have cinnamon because of its associative value and/or the symbol for Mercury painted in vermillion.

Of course, Mercury has long been used in dental amalgams for fillings and in thermometers; however, because of its toxic qualities it is being phased out of both these uses. Still, Lord Merku can help you take the temperature of things. He is very sensitive to dangerous situations and knows when things are beginning to get too hot and will warn you ahead of time. He will also help you to learn how to easily ad-lib in situations, to fill in where needed and to speak impromptu at the drop of a hat.

ॐॐॐ

We elves define an optimist as someone
With such good eyesight that they are able to see
Through the mists of doubt and illusion
To the radiance of Living Elfin
That awaits beyond, and in our dreams.

So many authors write about elves
As though we were their nightmares,
When we would rather be their dreams come true.

People often wonder how we elfin can still dance and sing when
Things are difficult and stress-filled.
And surely it is not that we don't see
And feel the suffering of the world.
It is just that we don't wish to participate in it.

CHAPTER THREE:

THE LORD AND LADIES OF
THE SOLID ELEMENTS,
THE MASTERS OF ENDURANCE
AND MANIFESTATION

Lithium, Lady Dongur (pronounced doan – goo-r)
Lady of Reform
Chant: "I would to tread a better way, not controlled
by life's wild fray."

Lithium is the lightest of metals and the least dense of solids and is very soft; soft enough to be cut with a knife, thus Lady Dongur is a very sensitive being and is the protector of all those who have been abused. When she appears it will probably be dressed in silver white or opalescent. However, she never appears naturally by herself. She always has protectors with her, often sea beings, sometimes, deep earth creatures or dwarves. By hers'elf she appears to have a soft luster, but since she frequently covers her being with mineral oil, she may shine.

Anyone who has been abused, such as battered women, abused children, or those who have been tortured all come under her protection and she can be fierce and volatile in her aid. If called to do so, she will react quickly and violently.

At the same time, she is also the patron/matron of all those who sincerely wish to transform themselves from abusers. Therefore, she looks after those who genuinely strive to overcome anger management issues, and even those recovering from drug and alcohol abuse, particularly if that abuse has lead them into violent behavior. If you know someone who is suffering from such abuse, or someone struggling to reform hir (his/her) own nature, then Lady Dongur can be of great assistance. If you are an abuser, you don't want to call on her, unless you have reformed or are determined to reform. And even then, she will ensure that you make up for the wrongs you have committed.

She is not fond of the elementals of gas generally, finding them to be too heady and intellectual for the most part. She is an emotional being and not inclined to chatter or to intellectualizing, which drive her crazy. If you bore her with too much talk, she may start throwing things at you or setting things on fire.

She has a lot of energy, thus the use of Lithium in batteries and thermonuclear devices. But she, as we said, also has her protective side, so Lithium is used in creating heat resistant glass and ceramics, and is also used as a lubricant. It is a great conductor of heat and electricity, thus whatever energy you invest into her you will get in return.

Lithium is also used as a mood-altering drug, particularly in treating depression, so once again you can see that Lady Dongur works to uplift those who have been downtrodden.

Lithium can float on water, thus Lady Dongur has province over ships and all things that float on the seas. She can teach you how to steady your emotions, how to direct them toward magical use without being overwhelmed by them, and how to rise above emotional situations without losing your sensitivity.

She reacts readily with water but not as much as the other alkali metals do. Depending on the amount of reaction you wish when calling her, whether you wish to get her volatile or to soothe her, you may have pure water (reactive) or salt water (soothing) in her triangle, or mineral oil if you really want her mellow and calm.

Since she was one of the three elements synthesized in the Big Bang, she is a very ancient being, so treat her with due respect. However, despite her great age, she is a rather rare being. Treat her well and she will do the same.

ॐॐॐ

Beryllium, Lady Sokyncor
(pronounced so – kin - core)
Lady of Allies
Chant: "Allies I do truly need, to strengthen me,
   my soul to feed."

Beryllium is another lightweight alkaline metal. Like Lady Dongur, Lady Sokyncor is a rare being, although not nearly as ancient as she. When making visible appearance, she will appear in steel gray or even in thin grayish armor. However, like the other alkaline metals, she is unlikely to appear alone. She may come wearing the gems beryl, aquamarine or emerald (both containing Beryl) and/or chrysoberyl. You may also wish to wear these or have them in your circle and/or magic triangle when summoning her.

While Beryllium is a brittle metal by itself, it makes a great ally, strengthening Aluminum, Copper, Iron and Nickel. And in these forms, it finds many uses in spacecraft, aircraft, and satellites. Therefore, she is a great matron of air and space travel in all its forms, including astral travel and out of the body experience. But her greatest power is in drawing appropriate allies who will assist, aid and strengthen the individual magician and help one achieve one's chosen visions and magical goals. If you need help rising in life, uplifting your spiritual and social station, she can be of tremendous help in finding the contacts and associations you need to elevate and illuminate yours'elf

51

and your life. Not only that, she will do this without causing increased karmic debt, or a significant disturbance in your life.

She can also take the heat and is usually unfazed in the most trying of circumstances. All the energy you put into her will be effectively and efficiently transformed into bringing about the results you desire. However, again, she doesn't like to be alone, and will rarely be found without companions. And, as we said, she prefers the company of the spirits of Aluminum, Copper, Iron and Nickel. Therefore, having any or all of these in your circle or triangle when conjuring her will surely aid in your success. Also, she prefers fresh water from streams to seawater, so note that fact in your conjurations. When she gets hot, she likes to hang out with the Oxygen twins. Some refer to her personality as both sweet and salty. So while she may be kind, she is also likely to swear like a sailor or make sexual innuendos or double-entendres.

Most x-rays go right through Beryllium without seeing it, so Lady Sokyncor is also the patron/matron of those who wish to travel the astral realms without being noticed or seen. She can render you invisible on those planes so you can observe or listen in without being detected or noticed.

Further, when in the company of her friends, this spirit has extremely high endurance. She doesn't give up, and will forge on despite any counter spells that are launched at her. If you wish to cut through a maze of confusing magic and counter spells, she, with her friends, is a good one to call.

Beryllium is also non-magnetic. While Lady Sokyncor likes to hang out with her friends, she is unlike to be overly influenced by them. She has her own mind and is quite determined in her quest. She gives them support and strength, flexibility and increased ability to endure difficult situations, and she will do the same for you. But she also has her own quest, which is to mirror your life so you can see yours'elf clearly and become more stable as an evolving spirit. She will help you endure.

ॐ ॐ ॐ

Boron, Lord Burnas (pronounced bew-r - nace)

Lord of Formation

Chant: "I would to build a better life, abundant and with
magic rife."

Boron is born of Cosmic Rays and thus represents
the Will of the Divine as it proceeds to create its
Vision through Nature. Boron is rare on the Earth
and in our Solar System, so Lord Burnas must be understood
to spend most of his time elsewhere. However, he does have
some connections here and will come if summoned, although it
may take a bit of time. He often travels via meteors. In its pure
form, Boron is a metalloid, which is to say partly metal, thus
Lord Burnas can be seen as an emissary or ambassador to the
metal elementals.

On Earth, he functions often through salts such as Borax,
and has a variety of skills including cleansing and purifying. He
has an easy relationship with water. And is a patron of the
country Turkey. So putting anything that is representative of
the Turkish culture in your circle can be very appealing to him
(he loves Turkish coffee). He also has a certain fondness for
the deserts of California. He will most likely appear wearing
brown or black or as a being with that coloration. He is a
patron of Brownies and of the Dark Elven.

His chief power, however, is as a builder. His Cosmic
task is to help fulfill the perfection and formation of the Divine
Magic through the Universe and upon the Earth by working
through the powers of Nature. He lends strength to all that he
builds so that it endures.

Boron has a toxicity level approximately equivalent to table
salt, so he is not a danger to you unless you call him too much,
at which point he may become testy at being overly bothered.
So if you summon him, tell him what you wish and trust that he
will fulfill your requests. You don't need to remind him. He
knows what he's doing and will get to it as soon as he can.

Because of his salty disposition, having salt water in your circle or triangle is conducive to his appearance. However, pure water has the effect of helping him to concentrate and focus. It somehow serves as a mirror into our world for him, so if you have a bowl of purified water in your triangle you may see him reflected there or arising from it. However, know that this elemental is in high demand, for he can help you build your dreams into reality, so don't waste his time. Tell him what you need and get on with your life.

While he has low toxicity to most animals he is, however, toxic to arthropods, so he's not fond of spiders, scorpions, millipedes, or centipedes so have no representations of these creatures in your circle or the room where you evoke. If you should have a tattoo of a scorpion, for instance, be sure to have it covered. On the other hand, if you are plagued by these beings, he is likely to drive them away, although except under extreme circumstances that is a waste of his magical powers. On the other hand, Boron is essential to life and, in the form of compounds, has a strengthening role in the cell walls of all plants and in enriching soils. And know that Lord Burnas does all this with very little. He is extremely efficient.

Carbon, Lord Benas (pronounced bee - nace)
Lord of Acquisition
Chant: "What I need will now appear, manifest what
      I hold dear."

Lord Benas is one of the few elementals that introduced himself to humanity in antiquity. Many, if not most, of the others hid from us as long as they could. Some surely are still hiding. He is the Lord of Acquisition. If what you desire is something material, he'll probably be able to get it for you. He knows nearly every other spirit and has a fairly friendly relationship with most of them, whether they are in high society or low or anywhere in between. He gets around and is welcomed nearly everywhere. Therefore, he can connected you to many other spirits. However, unlike Beryllium's Lady Sokyncor who will find the right associations for you, Lord Benas will connect you to almost any spirit you desire; whether they will help you or be a danger to you, will uplift you or tempt you, is another issue.

He knows the rich, the famous and the infamous. He knows society ladies, writers and coal miners and he tends to adapt himself to whatever company he keeps. So if he appears in response to your summoning, he will more than likely mirror your own culture and mode of dress, being as fancy or casual as you appear to be.

Carbon is the fourth most abundant element in the universe by mass following Hydrogen, Helium, and Oxygen. It is the 15th most abundant element in the Earth's crust and in the human body carbon is the second most abundant element after Oxygen, so this is a very influential and important spirit and while he has very real material powers, able as we said to get you almost anything, he is not without Cosmic connections. Carbon is, in fact, the chemical basis of all known life. While we list him here as an elemental, Lord Benas is pretty much a demi-God. While, because of his Cosmic power and influence, he could rise above Earthly concerns, he has little interest in doing so. He's a bit of an inventor. He likes to make things, to dabble, and most of all, as we say, to get you whatever it is you need. He is a patron of inventors, jerry riggers, and supply officers.

And Lord Benas is very much his own man/spirit. Others may attempt to influence or pressure him, but they are seldom successful. He does what he does for the love of doing it, and doesn't really care that much what others think or do. If he needs to be, he can be very hard.

Having diamonds, or even graphite or coal in your triangle or circle can be very helpful in summoning him, depending upon what it is you wish him to obtain for you. There is no direct cost for his service, but after you obtain what you desire you will most likely be approached by someone who needs something. Do your best to supply it if you can. Also, if you wish to make your request in the form of a requisition order, he will understand.

ॐॐॐ

Sodium, Lord Soulcor
(pronounced so – Yule - core)
Lord of the Laws of Hospitality
Chant: "Grant me sanctuary in your sacred magic aerie."

S odium is a highly reactive and soft metal thus Lord Soulcor is a very energetic spirit, who interacts actively with a lot of other spirits and is fairly adaptable. If summoned he will tend to appear in silvery white. He does not manifest on Earth by himself unless forced to do so but rather comes here through association with other elementals. He gets along well with Chlorine's Lady Lyrrata. Together they create the compound salt so necessary to life and so much a part of most magical traditions. Ruling the sixth most abundant element in the Earth's crust, Lord Soulcor is a very important being on Earth. Since Sodium can be found in feldspars, sodalite, rock salt and other minerals, these can be helpful in attracting him.

Besides having cleansing properties in coordination with Lady Lyrrata, as well as protective energies, salt is also noted and used as a token of commonality and of truce. To have someone accept your salt is to have them accept the protection of your hospitality and the responsibilities of a guest. This is an ancient magical formula and Lord Soulcor is the Master of the Laws of Hospitality and the ancient practices of sanctuary that are no longer practiced in much of the world but still hold sway among some of the tribal people of Afghanistan. If you are their guest, no harm will come to you and you, in turn, must offer no harm.

Therefore, Lord Soulcor is also the spirit that deals with the white flag of truce that is used in war for talks and of the principles of diplomatic immunity and by extension the establishment of embassies and thus ambassadors.

Sodium is a good conductor of electricity so the magic you invest in Lord Soulcor will be very effectively utilized.

If you have fire in your magic triangle Lord Soulcor is likely to appear wearing the color yellow. However, if you don't use flame he will appear as a lustrous silvery white that will darken the longer he manifests. Like his fellow alkali metal spirits, he's not fond of water and having water in your triangle is likely to piss him off. He may very well explode into a fit of anger. So caution there.

Lord Soulcor is a child of Carbon's Lord Benas, born of that being's incredible passion. In other words, born of Lord Benas' magic. Lord Soulcor also has an association with the planet Mercury, so associative perfumes, incenses, etc. in your circle can be of aid in your conjurations.

Sodium is often part of de-icing products, but again that merely reminds us of Lord Soulcor's diplomatic propensities. He seeks to de-ice Cold Wars to more friendly exchanges. If you have need of *breaking the ice* with someone or some group, he is a good spirit to call.

Magnesium, Lady Brisylcor
(pronounced bry – sill - core)
Lady of the Torch Bearers
Chant: "Light my way and show me clear the path to
all that I hold dear."

Lady Brisylcor will most likely appear wearing shimmering gray garments decorated with crystals. She is the matron of Torch Bearers. The Signal fires that went from Gondor to Rohan in the *Lord of the Rings*, would fall under her influence, as would the Olympic Torch. All those who have the job of lighting the way have her as a patron/matron. Of course, modern signal flares similarly come under her care. She is one of the alkali metals and thus holds great similarity and association with others of that group. They have their own little club you might say and all of them have a thing for crystals and crystalline structure.

This elemental rules the ninth most abundant element in the universe and the eighth most abundant element in the Earth's crust. It is also the fourth most common element in the Earth behind Iron, Oxygen and Silicon, so this is a very influential spirit, particularly here on Earth but also out in the Cosmos. After Sodium and Chlorine, Magnesium is the most abundant element dissolved in seawater, so you can see she has great influence there as well. Magnesium is used to strengthen Aluminum, producing a strong but lightweight metal, so there is a bond between Lady Brisylcor and Lady Nasnalney, the spirit of Aluminum.

Magnesium is also an important ingredient, we might say, in the human body, and is vital to all cells in the body. While it is only 11th by mass in the body, none-the-less, you can see that Lady Brisylcor has some influence over humans. So if you would like to sway someone toward you or your way of thinking, she can be of great help, particularly if you are endeavoring to warn or enlighten the individual. She is particularly good at soothing irritated people, sometimes even helping them to purge those things that are not really good for them, or are no longer of value in their lives. She encourages them to grow and prosper.

Some, however, feel she has a bit of a sour disposition herself or that she can be a bit tart. Having mineral water in your triangle can be of help in allaying this somewhat. If she seems a bit abrupt in response, ignore that, don't be insulted, simply tell her what you wish and let her get on with it. While she may appear in a blaze of light, she is unlikely to explode in a fit of anger. She is not as easy to upset as some of the others. She is very fond of fireworks, so if you have a sparkler going in your triangle, it will be very attractive to her.

You should know that once you set her to a task, she is very difficult to stop. So make sure you know exactly what you want her to do, because if you change your mind or want to adjust your request, she is unlikely to hear you as she is already ablaze with the magic you've evoked.

చేఁచేఁచేఁ

Aluminum, Lady Nasnalney
(pronounced nace – nail - neigh)
Lady of Ascension

Chant: "I shall rise above the Earth and to my soul I shall give birth."

---

ady Nasnalney tends to favor silvery white in her appearance, like many of her kindred. She rules aluminum which is a soft, non-magnetic and ductile metal and which is the third most abundant element in the Earth's crust, so she is a being with profound influence on Earth, and has the power of being the most abundant metal. She is highly reactive and so doesn't tend to appear alone, being an extreme extrovert and unlikely to be by herself for very long.

She has reflective powers, the ability to mirror the light of others, and in this way is a matron of psychiatrists and psychologists. But her great power is in the ability to help individuals ascend to a higher level of spiritual or social being, while helping them to resist the corrosive temptation to slip back into the habits and thought forms that kept them down in the first place. Therefore, she not only fosters air travel and is helpful to all who fly, invent, design or build planes, but she also aids any who seek to overcome addictions of any and every sort. She helps individuals to integrate, ground (feel comfortable and stabilize) and balance into a new and higher level of being.

This spirit is also extremely adaptable and flexible and thus can help both actors and those who perform yoga or who need to fit into situations that are new and unusual. She can help you endure the tensions and pressures of life without breaking down or losing your essential sense of s'elfhood.

Because Aluminum is a good conductor of heat and electricity, Lady Nasnalney is very efficient about using the energy you give her to help fulfill your desires. However, since salt does have a corrosive effect on Aluminum, calling Lady Nasnalney with Chlorine's Lady Lyrrata and Sodium's Lord Soulcor is not a good idea, since they tempt her to stray from her assigned task and degrade her magic and, of course, you don't want to call her with Mercury's Lord Merku who gives her an inferiority complex and denigrates her constantly, thinking of her as "cheap" and without gravitas.

However, since Aluminum in a stable form is created when Hydrogen fuses with Magnesium (in large stars or in supernovae), these spirits can be very supportive of her efforts. She also has a love of Feldspar so having that in your triangle can be conducive to her appearance.

Silicon, Lord Wiclor (pronounced wick - lore)

Lord of Mass Movements

Chant: "Gather now the common folk and let us cast away this yoke."

Silicon is a metalloid; therefore, Lord Wiclor can serve as an ambassador to most metal elementals. Ruling the eighth most common element in the universe by mass, it is an amazingly powerful being, however, he rarely appears as the pure free element in nature. Like many of the others, he likes company. Silicon can be found most often in dusts and sands of planetoids and planets in varieties of silicon dioxide also known as silica or silicates. Ninety percent of the Earth's crust is composed of silicates; therefore, silicon is the second most abundant element in the Earth's crust, with oxygen being first.

Lord Wiclor has province over mass movements, fads, and you might say, the ability for someone to become popular and accepted by the masses. He speaks of the power of the people, the wee little folk, to affect the great and powerful. If you would be famous, gain popularity, or affect mass movements, this is the spirit to call. He also rules both Democracy and Communism in their ideal manifestations.

Of course, mass movements can, like automobiles, be used for good or ill. Consider the effects that your magic will bring about so that you will not become unduly indebted by your actions. Still, with this spirit you can influence public opinion and sway it toward the just and the good, if that is your inclination. You can enlighten, uplift and make a better world for all of us.

Silicon is noted for its use in building materials, so you can help build a new world. It is also used in semiconductors in computers, so you can also influence thought forms around the world. Lord Wiclor will aid you in whatever you desire, it is you who must decide your moral and ethical boundaries and limitations.

Silicon is additionally the source of silicone, most popularly noted perhaps for its use in enlarging women's breasts. However, this again only speaks to the influence of popular notions, for women wouldn't bother to enlarge their breasts if it weren't for social pressures that tell them they'd be more popular for doing so.

Lord Wiclor's power also extends over sea sponges as well as being deposited in plant tissues including, interestingly, the silica cells and silicified trichomes (hairs) of Cannabis sativa, an incredibly versatile plant spirit who is used as a source of industrial fiber, thus paper and clothing, also building materials, as well as seed oil, as food, and for recreation, and additionally for spiritual and magical purposes as well as medicine. Thus Lord Wiclor can be said to have a very positive alliance with that being.

At the same time, Lord Wiclor can help you avoid mass movements, be free of such persuasion, to think for yours'elf and not merely accept peer pressure and what is commonly seen as being *politically correct* but to think clearly and analytically about what is happening in the world around you.

Certainly, if you summon him, you couldn't do much better than having a bowl of beach sand in your magic triangle.

Phosphorus, Lord Salfarsey
(pronounced sail – fair - say)
Lord of Auras

Chant: "Enlighten me and let me see the truth that will
my mind set free."

Lord Salfarsey is another extremely active extrovert and is never to be found alone on the earth. He always has an entourage with him. He will most likely appear in white or red or both. He is associated with the Roman god Lucifer, the light bearer or light bringer, who was seen by them as the God of the planet Venus, although he also has associations with the planet Mercury, therefore you may use correspondences to either or both of these in your magic circle. He further has association with the Greek god Prometheus who incurred the wrath of the other gods for teaching mankind how to create fire. His symbol would include a lightning bolt.

There are clear similarities between Lord Salfarsey and Magnesium's Lady Brisylcor; however, while the latter is really a light/torch bearer, Lord Salfarsey is more of a light bringer. He seeks to enlighten the enchanter and to increase one's glow or aura, the power of one's radiance.

Phosphorus is essential to life and so this is an important spirit who has influence in fertilizers, detergents, pesticides, and even nerve agents. Lord Salfarsey brings illumination; once again, however, how we use that increased knowledge, for good or ill, is up to us.

Curiously, his white form is the least stable and most reactive and volatile, so be cautious if he shows up only wearing white. He could be in a bit of a mood. However, he tends to mellow with time, so if you can keep his attention long enough the white will appear a bit yellowish and then gradually turn red. He's happier when he's red. This can be accentuated if you have a lot of light directed upon the magic triangle or have fire or heat in the triangle or make the room very warm or evoke

63

him on a hot day. Eventually, he will appear to be violet and in time and under a good deal of pressure, black, which is his most stable form. Of course, it all depends upon what you wish from him. If you wish him to go forth and do something immediately, probably the white and red forms are best, but if you wish him to stay and enlighten you, the violet and black forms are better. It is a matter of catching him in the right mood or waiting for it to develop. Our mother used to tell us not to ask anything of our father before he had his coffee. This is a similar idea.

Lord Salfarsey also has a fondness for Apatite, a gem that clears the mental centers and helps elevate the emotional centers. He further has association with the cultures of China, Russia, and Morocco, as well as having a particular fondness for Florida, Idaho, Tennessee, and Utah, surprising as that may be. Phosphorus is also commonly found in urine; however, we don't suggest you have a bowl of that in your magic triangle. On the other hand, it does tell us that Lord Salfarsey's advice, like most advice given in the world, is wasted. So don't call this spirit unless you are really willing to heed his hints and to change your life.

Sulfur, Lord Sulfur (pronounced sule – few-r)
Lord of the Crucible
Chant: "I would to change and from me rid, mys'elf
of those things I have hid."

While Phosphorus' Lord Salfarsey is related to the ancient Roman god Lucifer, Lord Sulfur is associated with Satan. They are, of course, commonly confused with each other or seen as one being, however, they represent two ancient Annunaki Lords, who were the brothers Enki (Lucifer) and Enlil (Satan). Satan is commonly viewed as being the devil or the Lord of Hell and this association, in part, surely comes from the fact that Sulfur is frequently found in volcanoes, which in the past would have been seen as eruptions of the fires and flames from the underworld of Hell. Sulfur is often referred to in some fundamentalist religious circles as Brimstone.

Lord Sulfur is seen by the elven as a Lord of Karma. It is his duty to oversee all those soulful spirits who have strayed so far from their true path that they need to be reformed in the Crucible of life, called in Alchemy: the anthanor. It was also referred to as the Philosophical furnace, sometimes the Furnace of Arcana, or even, the Tower furnace (see the major arcana Tarot card #16, the Tower). These beings simply cannot go on without burning away all that stands between them and soulful association with others. They come to be reborn from this experience, however, it is surely a painful process, and is thus seen at times as being hellish. The expression *what doesn't kill you makes you stronger* can be associated with his efforts. This, of course, means that he is not the Lord of Hell (which doesn't really exist), torturing errant souls forever, but the Lord of karmic purgatory.

If summoned, Lord Sulfur will appear wearing bright yellow. Unlike Lord Salfarsey, Lord Sulfur can be found alone and may very well come without any attendants. However, Lord Sulfur has a very powerful odor, commonly associated with rotten eggs, which is in fact the source of the skunk's fragrant aroma. But it is also the source for grapefruit's smell as well as the odor of garlic. We suggest you use grapefruit or garlic as an offering to this spirit. He has a fondness for hot springs and, of course, all places that are near volcanoes, active, dormant or extinct, so warm mineral water can be used in the triangle. He is also associated with Jupiter's volcanic moon Io, so Jupiterian correspondences may be utilized in your circle, although his nature is somewhat closer to Saturn's.

Like Lord Salfarsey, Lord Sulfur has a powerful influence on the life of plants, often being used as a fertilizer. Sulfur was also used in the past for making quality gunpowder, this being the source of the acrid smell of battlefields. General Sherman is quoted as saying, "War is Hell", and we would add *and smells like it, too.* Sulfur is essential to all life forms and so Lord Sulfur is a very powerful being. It is found in a number of vitamins, in particular thiamine (vitamin B₁), whose name comes from the Greek word for Sulfur.

If you are enduring an extremely trying time in your life, suffering things that seem almost unbearable, then Lord Sulfur can help you find the meaning in your suffering, help you endure through it and also help you to gain from it. He will not take it away from you, after all it is his job to see you go through it, but he will help you grow stronger from it.

꙳꙳꙳

Potassium, Lady Nårhemåcor
(pronounced nar – he – mah - core)
Lady of Soulful Connections
Chant: "I would you feel for me as true as I do feel so
        much for you."

While Lady Nårhemåcor will appear as wearing silvery white, like many of her kin, she can be distinguished by the lilac aura around her. She is in many ways similar to Sodium's Lord Soulcor, the Master of Hospitality; however, Lady Nårhemåcor deals with the development of soulful connection, or simpatico or fellow feeling. She awakens feelings of friendship and kinship between people. If you wish someone to like you better, then this is the spirit to evoke.

Potassium is highly reactive to water so while she rules the development of the feeling connection between people, she doesn't like the overly sentimental or emotional. She hates trauma drama and overly dramatic individuals. Neither does she care for those who are syrupy or overly sweet, whom she feels to be false. She is looking to develop genuine and sincere connections between people. She is somewhat salty as a spirit and therefore she prefers people who are rough but real to those who pretend to be refined but are artificial in their relationships.

This spirit has a fondness for seawater and mineral water, either or both of which can be used in the magic triangle to evoke her.

Potassium ions are vital in the proper functioning of all living cells, and are a key aspect in nerve transmission. If potassium is depleted in animals, which includes humans, it may result in number of cardiac dysfunctions. In other words, Lady Nårhemåcor is vital to the proper functioning of the heart and of heartfelt connections and the communication of feelings. She makes it so that one not only feels the connection but that the sense of connection is effectively communicated and shared.

Potassium can be found commonly in fresh fruits and vegetables, so Lady Nårhemåcor is a vegetarian. If you wish to make an offering to her, fresh fruits and vegetables are a good choice. Potash, from which the name Potassium is derived, is sacred to her and can be used in the magic triangle. Potash is essentially plant ashes soaked in water that form coppery colored crystals. It is mostly used in the world for fertilizers, which tells us that Lady Nårhemåcor helps enrich our lives and helps relationships to grow.

Representing a soft metal that can be cut with a knife, Lady Nårhemåcor is very sensitive. Courtesy and manners are important to her. Treat her with due consideration. She's really a romantic. However, she is also sensitive in terms of picking up psychic vibrations so, as we said, she can spot a phony a mile off. You need to genuinely care about the person you wish to connect with and your treatment of Lady Nårhemåcor must be sincere. And she doesn't much care for spirits of water and Oxygen together. If she encounters them she can get volatile quickly. In other words, she doesn't care for blowhards or for emotional chitter-chatter, or those who say I love you but don't really mean it. Those who go on and on about their relationships without actually feeling anything genuine, drive her crazy, so once again, sincerity is the key. She hates those who sweet talk others or lie to them with nothing real behind their words. She believes in loving action not talk.

დ დ დ

Calcium, Lady Cahyrcor
(pronounced cah - her - core)
Lady of Supporting Structures
Chant: "I would to link to others true, support to give
for all we do."

Calcium is another alkaline earth metal and is fifth, by mass, in abundance in the Earth's crust, thus Lady Cahyrcor is another powerful being. It is also the fifth most abundant element in seawater and in the human body. Because of this association with the number five, the pentacle works well with her. She will tend to appear wearing gray, although she may very well have a brick red or orange red aura, particularly if summoned in warm atmospheres. However, being extremely reactive, extroverted Lady Cahyrcor almost never appears alone. She loves to associate.

This element is perhaps best known for its part in the mineralization of bone, teeth and shells, therefore, this elemental rules support structures. The name Calcium comes from a Latin word for lime, which is noted for its use in building materials. If you are alone, for which she has great sympathy, and have need for a support structure, she will surely aid you. Also, if you wish to create a business, coven or organization of any kind, whether its intention is to function publicly or privately, she can be of help there as well.

While Calcium is a soft metal, it can be cut with a knife, but with difficulty. So we might say that while Lady Cahyrcor is a bit of a softy, she's no pushover. She can be somewhat ponderous and difficult to get going, but if you get her aroused she will remain so for a long time. So first you must plead your case and catch her interest. If she agrees to help you, you can be sure she will endure until the magic is done. If you have water in the triangle, this will tend to increase her rate of reaction.

Of course, if you, like many magic wielders, have a fondness for skulls, bones, shells, and teeth of various creatures that you have found, you can put these in the magic triangle, for she is also fond of these things. The minerals calcite, dolomite and gypsum may also be used to attract her. She is the matron/patron of the Tooth Fairy.

ॐॐॐ

Scandium, Lord Nolcor (pronounced nole - core)
Lord of Assistants
Chant: "Bring to me a trusted one, to aid me till the
     magic's done."

Lord Nolcor loves Scandinavia and all things from that culture, although he also has a fondness for Madagascar and probably vacations there in the winter months. He will appear in silvery-white metallic garb and will have a yellowish or pinkish aura, and is best noted as an ally of Aluminum's Lady Nasnalney, thus this spirit rules assistants, secretaries, vice-presidents, chief-of-staffs and all those who serve as the primary assistant for others, also of nurses, dental assistants, master sergeants and flight attendants and copilots. We've heard it said that if you wish something to get done you need to go strait to the top, but often you have to get by their executive assistant first. So if you would influence such a person, or draw one to you to aid you in your magic, Lord Nolcor can help with this.

Scandium is listed as a rare earth element, although in fact, rare earth elements are actually quite plentiful on the Earth. It is rather like the fact that assistants, nurses and secretaries don't get much credit or have high social status while doing most of the work. They aren't noticed much, but they are the worker organization bees of the world.

While Scandium is only the 50th most common element on Earth and the 35th most abundant in the Earth's crust, it is the 23rd most common element to be found in the Sun, so while Lord Nolcor is not the most powerful of spirits, he has his influence and that influence is greatest in relationship to the Sun, which is to say, the boss.

Lord Nolcor has relationship with the minerals thortveitite, euxenite, and gadolinite but these are rare and hard to come by although euxenite has been known to be used in jewelry. However, he has also been known to like apatite so that would be the gem to use in calling him. This gem has an association with the planet Mercury and its correspondences.

Scandium has low to moderate toxicity. Thus remember that Lord Nolcor needs to be approached in the right fashion. Secretaries can be helpful, even friendly if you approach them the right way, but if you are not careful and respectful they will work against you, blocking your way; but that is your own fault. Also, Scandium is often found in the company of others, particularly the other rare earth elementals, and has even been known to hang with the very dangerous Uranium spirits. Keep that in mind.

So again, treat him with respect and remember he can be the key to obtaining what you wish from other spirits. But even though he is a means to an end, don't take him for granted.

> ## Titanium, Lord Jalcor (pronounced jail - core)
> Lord of Righteousness
> Chant: "I've done what's right and now I would receive
> the judgment of the good."

*L*ord Jalcor will appear in silver with a white aura. He is a very powerful and elevated being and is not easily influenced, persuaded or tempted by any other spirit. Therefore, he is seen as the Lord of the Righteous. Not the self-righteous, mind you, but a representative of what is truly right, fair and just. If you feel you have been treated unfairly and are in the right, this is the spirit to call. He has province over laws and courts on all levels of being, including the Court of the Stars, the high court of Karmic Justice.

Titanium is named for the Titans of Greek mythology, those powerful beings who were the second order of Divine manifestation and ruled prior to the ascension of the Olympian Gods. Among the Titans were the 12 children of the Earth Mother Gaia and the Sky Father Uranus. So while Lord Jalcor has strong cosmic connections, he is very attached to the evolutionary development on Earth. And he is found nearly everywhere on the Earth, so his influence is truly widespread. He loves the Earth as a whole.

Titanium is the ninth most abundant element in Earth's crust and it is the seventh most abundant metal, so Lord Jalcor is a being of some significance on the Earth. And while he travels worldwide, he does have an especial love for the cultures and peoples of Australia, Canada, China, India, Mozambique, New Zealand, Norway, South Africa and the Ukraine.

He is associated with the minerals rutile and ilmenite, although rutile quartz is probably the one you wish for evoking him. It is a mineral used for magnetic balancing and healing of the individual's body and soul. Therefore, Lord Jalcor will bring balance and fairness to any situation you find yours'elf within.

This spirit has an easy association with the spirits of Iron, Aluminum, Vanadium, and Molybdenum and together they accomplish many things, so Lord Jalcor has a great many talents and abilities but, again, all of those stem from his primary aspect of being incredibly strong and resistant to the corrosive influence of others. He is powerful without being ponderous, keeping a light heart and an enlightened view in all situations. Titanium is non-magnetic and also a poor conductor of heat and electricity, which is to say, Lord Jalcor will arrive at his own opinion once he analyzes a situation and cannot be influenced or persuaded by anything but the truth.

However, he is not at his best in extreme heat, so keep your magic room cool or at least moderate when summoning him.

He does get along well and forms a quick association with the spirits of Oxygen, who tend to be protective of him. Titanium however, unlike many of the previous spirits is not influenced much by water or air, thus once again, neither emotion nor spurious arguments will affect him. Neither weeping nor rhetoric moves him.

Lord Jalcor does, however, get fired up (some say enflamed) by the spirits of Nitrogen, so keep that in mind.

శురుశుశు

Vanadium, Lady Afrocor (pronounced a - fro - core)
Lady of Catalytic Action
Chant: "I would it start and quickly now, stir the cauldron, and show me how."

Appearing in bluish silvery gray, Lady Afrocor has a powerful catalytic influence. If you have done all that you can to bring the right elements together and things still don't seem to be going in your life, then she will stir them up and set them into action. If developments have been moving too slowly, she will accelerate them. She is an incredibly adaptable spirit and has a great deal of power because of this. However, she is unlikely to appear alone and usually will come with Oxygen's Lady Emper, at the very least, if not others.

She has a love of Mexico and its people and culture and knows a great deal about Nagualism and Sorcery. Interestingly, despite her tendency to appear in bluish silvery gray, she has a love for the color red, so decorating the magic room with red can be helpful in attracting her. And she will most likely manifest as a young beautiful woman with rosy cheeks, vibrant and filled with life. Sometimes she will wear a multi-colored, rainbow or tie-dyed robe. She also has a fondness for China, South Africa and Russia and their peoples, cultures and magical traditions.

However, note that she has some association with those dangerous Uranium spirits and has a low to moderate toxicity herself. She can also be found hanging out with the Iron spirits, but that is only a problem if you believe that Iron is dangerous to faerie folk. Still be cautious with her, you may get more than you bargained for. As they say, *be careful what you wish for*, and *don't bite off more than you can chew*. She will get things going, but are you really ready for that to happen?

However, that being said, she has the great ability of being able to get cynical, acidic, and negative people directed toward positive goals, although they'll still be cynical and negative about it. She is mostly indifferent to their corrosive attitudes and she is a very tough spirit. If you have such naysayers around, she is not beyond stirring them up to help fulfill your magical desires. She, actually, finds this amusing. She knows that deep within they are only so very negative because they really want to believe.

And, unlike many of the others mentioned previously, she doesn't have a problem with water, in fact, she has a fondness for the sea. She also has an association with the magnetic mineral magnetite, which she uses to help those around her to focus, so you can put this in your triangle of summoning. She also is attracted to vanadinite, but if you can't find this mineral know that it is one of the apatite group, so some other form of apatite can be used instead.

ॐ ॐ ॐ

Chromium, Lady Luharcor
(pronounced lou - hair - core)
Lady of Confidence
Chant: "My mind my own to wield and see, the world
as it does best suit me."

Lady Luharcor is a tough spirit, who will appear wearing shiny steel grey armor that will reflect the light and the colors of the room she is called within. She has a very positive sense of herself, has her own spin on life, and can be of service to anyone who needs to develop hir (his/her) own sense of s'elf worth. If you have an inferiority complex, or feel less than competent in any area, she will aid you to gain a positive outlook and attitude that will enable you to succeed at whatever you attempt. Chromium is magnetic, so you will certainly attract others by being true to yours'elf; however, you will find only certain individuals will be attracted. Trust that these are the right ones for you, although, a certain ordering may be needed, which is to say, you may need to keep certain individuals at a greater distance than others.

Remember, however, in large amounts Chromium can be toxic and carcinogenic. So it is important that you retain a sense of humility and modesty in dealing with others, for if left unbalanced one could easily become arrogant by following Lady Luharcor's advice. Feel good about yours'elf but don't overdo it. She will make you feel so good that you may miss the importance of remembering that everyone else wishes to feel the same way about thems'elves. Remember that your way is the right way for you but not necessarily the way for everyone.

This spirit is very smart, sharp, and nearly incorruptible, which is to say, very unmoved by the opinions and particularly criticism of others. She really doesn't care about other people's opinions concerning her and can help you become impervious to gossip, idle backstabbing, and negative comments.

While she may appear alone, she is most often found in the company of the spirits of Oxygen and Iron. Together they are a very tough group of spirits. And she has a love of color and the colorful, so dress and/or decorate the magic room with as much and as many colors as you like.

This spirit is probably best known for her influence in creating stainless steel and even more so for her influence in creating Chrome. She is the Queen of Chrome. So having anything chrome in the triangle or magic circle is very attractive to her. If you have any chrome magic tools, it would be greatly helpful. If you don't have chrome, stainless steel will work as well. Use your stainless steel silverware if you need to.

Lady Luharcor has a love of South Africa, Kazakhstan, India, Russia, and Turkey and their peoples, magic and cultures. She also has association with the crystal crocoite, and for quartz in general and also for hematite. Crocoite is noted for healing back pain, thus can help one get a backbone, that is to say, to stand up for ones'elf, while hematite helps one reevaluate ones'elf. Crocoite is noted for being used for creating yellow pigment in paints; yellow being the color of balance. Lady Luharcor also has a love of rubies and will probably appear wearing one or more of these gems. Rudy helps one become more generous and willing to be of service to others, so that makes a nice balancing aspect for the primary individualistic thrust of this spirit's influence.

๛๛๛

Manganese, Lady Morten (pronounced more - teen)

Lady of Style

Chant: "I embrace my style unique, and those attracted to my freak."

Manganese's Lady Morten has a great many similarities to Chromium's Lady Luharcor, however, while Lady Luharcor gives one confidence in ones'elf, Lady Morten aids one to become confident in one's own style, life-style, individual style but also and particularly one's artistic style and will help you individuate so you are an increasingly unique and vibrant artist.

Lady Morten tends to hang out with the Iron spirits, so traditional faery folk should know this, although she has association with other minerals as well. She will never appear alone, being another extrovert who just loves company. She will manifest in a shiny silvery form or wearing such garb although it will be darker than Lady Luharcor's raiment, more nacreous than brightly lustrous.

She has a very important part in the life of plants and animals but only in trace amounts. You don't want to call her too much and you wish to be brief. Her influence is powerful, but while developing and embracing your own style is very important, you also may wish to consider your audience, unless you are one of those writers or artists who do their art only for their own satisfaction and don't want anyone else to see it. It's not about selling out for fame but finding your niche where those who really understand your work will be have the opportunity to express their appreciation. You don't wish, we expect, to believe in your art to the point where you are no longer willing to develop and evolve as an artist. So while Lady Morten will give you confidence in your style, don't let her persuade you to listen only to the obsequious. Believe in your style yet always continue to improve what you are doing.

Lady Morten will surely attract an audience to you. Your work will develop a magnetic aspect that will harmonize with others of like kind, if that is your desire.

She is further a patron/matron of glass workers of every kind, those who work in glass blowing, but also those who do stained glass, and even those who merely create clear glass windows, doors and glassware. She loves glass. If you use a cut glass chalice in your circle, she will find this very appealing. She also has affection for Greece and the culture of the Greek people but especially a love of ancient Greece and Greek mythology. But she also has associations with South Africa, Brazil, Ukraine, Australia, India, China, and Gabon, and the deep, deep ocean.

Like Lady Luharcor, you can use stainless steel in your triangle or magic circle, but also the alloy spiegeleisen that has a shimmering rainbow quality to it. Lady Morten further has a positive association with the Aluminum spirits so you can have Aluminum in your triangle as well.

And since Manganese is 12th in line in composition of the Earth's crust, this is a fairly important spirit. She does a lot with very little.

Curiously, she also has the power to calm and quiet ghosts and poltergeists, as well as silence those who would criticize and knock your work. She holds power over Tommy-knockers.

Remember, she is best in small doses. Too much association with her often causes people to lose their perspective on life and sometimes to even develop a warped point of view so caution is advised.

Iron, Lord Feri (pronounced fee - rye)
Lord of Discipline

Chant: "Make me strong and power filled, limited by what I've willed."

Iron has long been said to be poisonous to the elven and faerie folk, weakening us in the same way Kryptonite weakens Superman. As we pointed out in the section on Mercury's Lord Merku this is most probably due to the fact that Iron is used to hold or bind Quicksilver/Mercury, an element with which our people are frequently associated. Lord Feri does not harm the elven, or most of us anyway, but can limit us and thus he is the Master of S'elf Discipline, limitation and the development of Will Power. Faerie elf folk are not generally inclined towards those who attempt to limit us. We instinctually question authority. However, the I Ching states that we shape our lives and gain significance as free spirits by voluntarily accepting the limitations that duty places upon us and many magical traditions define magic as creating change in conformity to one's will. It is that *will* that Lord Feri seeks to empower and increase in us by helping us to develop s'elf discipline.

By mass, Iron is the most common element on Earth, therefore, this is a very important spirit. It is the fourth most common element in the Earth's crust. Lord Feri has great power here. It is also the sixth most abundant element in the Universe, so his influence isn't restricted to the Earth. He will tend to manifest in a lustrous silver grey having a rust colored aura. While he can be a bit soft himself, he is greatly strengthened by association with the spirits of Carbon. So don't think you are dealing with someone so tough that you can't relate to him. He can help you develop your will power because he's been through that process himself. He's not unsympathetic to the weak, the procrastinators and the vacillators.

He was known to the ancients, although not as well known as the copper spirits, but he definitely has a long associations with wizards and warriors. Curiously, while others might advise you to rid yours'elf of impurities, he will show you how to turn your weaknesses, peculiarities and eccentricities into strengths, to take that shadow element of yours'elf, integrate it and use it toward the fulfillment of your goals.

Of course, Iron has an important part in the lives of plants and animals and is noted for its strengthening of the blood. It is important in the development of red blood cells. Being red blooded is an analogy for being strong and robust.

While Iron is not magnetic in and of itself, it can become magnetic, so while Lord Feri will help you develop your will power through discipline he will also help you remain open to the influence and help of others so you don't become inflexible or lose your ability to adapt.

Iron, particularly cast Iron, is pretty easy to find so you can use that, or even anything steel in your triangle to attract Lord Feri. Also hematite and magnetite can be used.

ॐॐॐ

Cobalt, Lady Korsal (pronounced core - sail)
Lady of Loyalty
Chant: "Bring to me the ones so true, together we may great things do."

Lady Korsal, while she appears in shiny silver grey, is noted for her beautiful blue aura. She has province over Kobolds (from whom the name Cobalt comes) and Goblins. She rules the development of loyalty and if you attain her aid as a spirit you can be sure she will be totally faithful to whatever agreement you arrange between you. Unless forced to do so, she will never appear alone and instead will come with her friends or her gang you might say, mostly

79

likely the Iron, Copper or Nickel spirits. The Marine Corps motto Semper Fidelis meaning *always faithful* or *always loyal* aptly describes the power of this spirit and thus she has province over the marines as well. She is particularly inclined toward the region in Africa called the Congo and also the nation state of Zambia. She further has a love of the South Pacific island and peoples of New Caledonia and of the province of Ontario in Canada. She further has affection for Norway, Sweden, Saxony (in Germany) and Hungary, particularly a love of their ancient heritage.

She is noted for the production of blue pigment used in paints, glass, and other products. She is true blue, genuine, real and faithful to her own people and those with whom she bonds. If you have need of attracting those who will be loyal to you or your vision, or have need of developing loyalty in your own character, she is a good spirit to evoke.

Cobalt has an active part in vitamin B_{12} which is so important to the functioning of the brain and the nervous system. It plays a central role in coenzymes. It is also an important nutrient for algae, fungus and bacteria. So this spirit can play an important role in developing a reliable coven or group. She helps feed the positive aspects that keep communication open and the individuals involved dedicated to the common cause. However, Cobalt ingestion is toxic other than in the minutest forms, so keep in mind that loyalty can lead to fanaticism, and that a fanatical demand of loyalty can drive people away from you, destroying the atmosphere you wish to create.

However, Lady Korsal doesn't get along well with the elements Fluorine, Chlorine, Bromine, Iodine, and Astatine, which are halogens, or with the Sulfur spirits. They don't like her and will undo her magic. She is pretty much indifferent to the Hydrogen and Nitrogen spirits and almost never associates with them.

Naturally, one can use the rock Cobalt or Cobalt glass to attract her. These have a balancing and elevating effect upon her.

๛๛๛

Nickel, Lady Maletil (pronounced may – lee - tile)
Lady of Personality
Chant: "Help me refine my image sweet attractive now to all I meet."

This spirit often appears in lustrous silvery white with a golden tinge. She has occasionally been mistaken for the spirit of Copper and more often for the spirit of Silver. She is both hard and adaptable, which seem to be opposites, but they are not. Those who are fittest are most adaptable as Darwin points out to us. She holds sway over the development of the personality, the persona, and also over advertising of all kinds and can help the conjuror present hirs'elf in such a way as to be found pleasing to those sHe (she/he) wishes to influence. She helps the enchanter make a good first impression but also helps hir to gradually align hir character and hir personality so that the positive being sHe projects is indeed the being sHe truly is. Eventually, *what you see is what you get* is the result of this effort.

There are many quite good and upright individuals who have difficulty in life because they do not know how to interact with others in a way that will prove successful. They are hampered by their personalities (or lack thereof). Lady Maletil will help the magician develop hir (his/her) personality so that it is truly a magical power that increases hir ability to obtain whatever sHe (she/he) desires and to obtain whatever help is needed in doing so.

This spirit has an association with dwarves, kobolds, tommy-knockers and other mining spirits, and in particular with mischievous sprites. She is often found with the spirits of Cobalt and Iron. And she, like Iron, has a strong association with extraterrestrial beings and particularly with the spirits that use meteors and meteorites as vehicles. Like Cobalt's Lady Korsal, she's fond of the Ontario region of Canada and the Pacific island of New Caledonia. They are frequently seen there together. She further has a fondness for the Russian people and culture and their shamanic traditions. She furthermore has a love of Australia and its peoples, and for the peoples and cultures of the Philippines and Indonesia. In the U.S.A., she can be found hanging out mostly in Oregon and Michigan.

Nickel has been commonly used in coins, which reminds us of this spirit's ability to help us obtain whatever it is we desire. However, her great power, as we said, is her ability to gain allies through the development of the personality. If you can make friends easily you have access to whatever you desire in the world. Lady Maletil can be very magnetic.

However, this spirit also has association with the Sulfur and Arsenic spirits, so she is not simply Miss Sweet (Lady Sweet you will find under Strontium). She knows how to get along with a variety of spirits and sway them to her will. She holds the power of personality.

ಌಌಌ

Copper, Lord Dwartyn (pronounced dware - tin)
Lord of Ancient Civilizations
Chant: "Tell me of wonders past, so I may with magic make them last."

The Lord of Copper, Lord Dwartyn, is an extremely adaptable being, ancient, beloved and related to spirits of Silver and Gold. In appearance, as you know, he has a red-orange look with a greenish aura that grows deeper and stronger the longer he manifests. He knows nearly everything about ancient civilizations, both known, mythologized and those long forgotten. He can tell you of Babylon and Ancient Egypt, of Caldea, India, China and many other places. He can inform you, if you wish, of Atlantis, Mu, Ys and of civilizations whose names and existence is no longer remembered except in the Collective Unconscious of those peoples who are directly or indirectly their descendants. He has an ancient home on the Island of Cyprus.

In modern times, Lord Dwartyn can often be found in Michigan. He can also be found in Utah and New Mexico and is sympathetic to the cultures of those who live there. Elsewhere in the world, he has a love of Chile and its people and cultures, particular the native folks and their ancient magical practices. In fact, he has a great love of all native peoples and their crafts. You can't do much better than decorating your magic room with Native American artifacts when calling him.

Copper is important to all living organisms in trace amounts as a dietary mineral. As Lord Dwartyn says *a little culture never hurt anyone* and *we can all learn from the lessons of the past*, if only to avoid the mistakes our ancestors made. In humanoids, Copper is found principally in the liver, muscle and bone. Knowledge of the past strengthens us and provides us with a cultural structure while telling us that the way things currently are established are not always the way they have ever been or need to be. The saying *those who ignore history tend to repeat it* is germane here.

Like many other metals, Copper was created in the stars, so Lord Dwartyn can speak not only of the ancient civilizations of the Earth but also of the energy and spirit civilizations of the stars, and thus he can tell you a great deal about ancient Faerie and the origins of the Star Children.

Naturally, in summoning Lord Dwartyn you can put anything that is copper in your magic circle or triangle, including copper pennies if you have them, bracelets, cups, or plates. Also, brass and bronze, which are copper alloys, can be used to attract him.

As is the case with Aluminum, Copper is 100% recyclable without any degradation of quality, thus Lord Dwartyn reminds us that whatever good there was in the past can always be recreated. To those who go on and on about *the good olde days*, he says to take the best of the past and make it live in the present. Although, it is true that many of those folks, like those who think that *the grass is greener on the other side of the hill*, usually don't know what they are talking about, and are viewing the past from the patina of history, legend and mythology, which in themselves often tend to reclaim the best of the past while ignoring the worst.

Copper is biostatic, a word that denotes that bacteria will not grow on it. Once again, Lord Dwartyn can tell you how to avoid the mistakes of the past while utilizing its values and create for yours'elf and the world a better place to be. At the same time, Copper can be used to ward off demons (see Mike Carey's great novel *The Naming of the Beasts*). Lord Dwartyn is a good spirit to know. Also, he will tell you that civilization is nothing without art, a truly faerie point of view.

Zinc, Lord Wiz (pronounced wise)
Lord of Childhood Education
Chant: "In the Beginning, let me see, the very best that I can be."

Zinc with Copper is used to create brass so Lord Wiz has a strong association with Lord Dwartyn. Lord Wiz is in charge of early education and development since Zinc is so very important in prenatal and postnatal care. Lord Wiz not only rules the early development of children but of novices in any and every field of learning and life. Anything you wish to learn can be aided with the assistance of this elemental spirit whether it is horseback riding, playing the piano, magic, chemistry or some sub-field thereof. Thus if, for instance, you are already an accomplished magician but don't know anything of Chaos Magick but would like to learn, Lord Wiz will surely aide you and lead you to the texts, information and experiences that would be most fruitful for you.

There are some indications that the noted German magician and alchemist Paracelsus named this element. Ancient alchemists would burn Zinc in the air to create a white powder they called Philosopher's Snow. Philosopher's Snow is an analogy for wisdom that is shed upon us when we evoke, make burnt offerings, to the spirits/gods. Burnt offerings were traditionally the smoke of animals offered to the deities, but also, as in Taoist calligraphic magic, the burning of inscribed paper talismans. These days, the backyard barbecues that are so prevalent would be equivalent to such animal offerings, if indeed the charcoals and fire starters were not poisonous, as they so often are, and the meat was actually offered to the spirits, which it is if the individuals say *grace* before consuming it.

This spirit is most likely to appear in shiny bluish-white and since it represents a diamagnetic metal, that is to say a metal that creates an induced magnetic field opposite to an outwardly applied magnetic field, what he urges you to learn are the principles rather than merely accepting the dogma of any field of study. He urges you to obtain the facts and to think for yours'elf about all things, challenging and testing the accepted points of view. Because Zinc has a hexagonal crystalline structure, the six-pointed star or Star of David works well when summoning this spirit, as well as the continuous line six-pointed star that is currently in vogue in many magic circles.

This spirit gets on very well and forms close and productive bonds with the spirits of Aluminum, Antimony, Bismuth, Gold, Iron, Lead, Mercury, Silver, Tin, Magnesium, Cobalt, Nickel, Tellurium and Sodium.

However, like many old fashion teachers, if you try his patience or attempt to get too much at once instead of following the procedures he recommends, he is likely to scream at you. He believes in step-by-step development, so don't get beyond yours'elf in your eagerness to advance. Remember, in calling this spirit, as with most other spirits, moderation and balance are the keys to success.

He is also known to have a great affection for the peoples and traditions of Australia, Canada and the United States and also for the Persian people.

ॐॐॐ

Gallium, Lady Romecor
(pronounced row – me - core)
Lady of Ambience

Chant: "I would to change the feeling here, a better vibe to make things clear."

Like most of the extroverted elementals, Lady Romecor will never appear without company. She is none-the-less a soft, shy being and rather than appearing at the front of the group, will probably be found hovering near the rear. You will have to summon her closer to speak with her and even then she is likely to peek out from behind some other spirit. She will nearly always dress in silvery colored garments, often emitting a bluish-violet aura. She has an association with the garnet family, so these stones can be used in attracting her and because garnets are sometimes referred to as a warrior stone, they will increase her strength and confidence. There is in fact a gadolinium gallium garnet that is artificially created for jewelry that is particularly associated with her.

Her great power is to be found in her ability to alter the atmosphere or ambience of any place or gathering. She can heat things up or calm them down or simply alter the feeling one gets according to your desire. Therefore, she can also be called to change the psychic atmosphere of some place you may sense doesn't feel quite right.

She also has the ability, like Lord Merku, to take the temperature of things, that is to sense the level of excitement or lack thereof concerning any place or event, and can let you know ahead of time how certain events will go over, and she is a lot easier to deal with than Lord Merku, who can be a bit dangerous when in one of his moods and, in general, she is more eco-friendly and concerned with the Earth than that more Cosmic being.

However, she is a quite sensitive being, as you might expect, so she appreciates the trouble taken to woo her with extra courtesy and refined behavior. If treated rudely she could become a bit sharp. You may wish to use a fine, decorative teacup as a chalice. On the other hand, you should know that she tends to get on the nerves of some metal elementals who can become quite brittle around her. They are often of a rougher sort and her refined being sets their teeth on edge. Simultaneously, she rather likes to hang around with them and in time will change their attitude. She also has the ability to change the attitude of others toward you. She will melt their hearts and make them feel the most incredible sympathy and compassion.

Remember though that she can be a slippery being, changing from one mood to another quite quickly, so if you get an agreement from her that satisfies you, conclude your ceremony before she changes her mind. Note that she has a particular fondness for the ancient peoples of Gaul or ancient Celtic France.

> ## Germanium, Lord Revancor
> ## (pronounced ree - vain - core)
> Lord of Discrimination
> Chant: "Bring to me those spirits right, let others pass
> on through the night."

Lord Revancor is most likely to appear as a young man in lustrous grayish white armor, looking very much like an ancient Germanic Teutonic knight. He has a long association with the Saxons. He also has a kinship to the peoples of Inner Mongolia. He is a cousin to the Tin and Silicon spirits. He is sometimes confused for the elemental of Silicon and at other times for Antimony. He is frequently found in the company of the Oxygen spirits, although he has also been known to hang out with the Silver, Zinc and Sulfur spirits. And occasionally with the Lead and Copper spirits. He is part of the Carbon spirit family and because of this Diamonds or Coal can be used in calling him.

He is best noted as the Lord of Discrimination, not prejudice, as that term is often used, but as a spirit that knows who is right for you as a developing enchanter and spiritual being and who would merely waste your time, or worse yet, hinder your progress. If you have already called a magnetic spirit to attract others to you, Lord Revancor will make sure that those who come to you are the best available to help you. For attraction magic, unless very specifically formulated, casts a wide spectrum that draws many without discrimination. Lord Revancor will hone that vibration to select the most fruitful relationships possible so your energy and magic will be increased and magnified.

However, Lord Revancor does not get along with water, so you would probably be better having wine or some other substance in your chalice, although due to his somewhat austere nature that holds little interest for him.

You should know that Lord Revancor's coming was foretold. He was the subject and fulfillment of prophecy and is well aware of this status, so remember to treat him with due respect as though you were speaking to some noble in a fantasy novel, or some lord in Shakespeare. He loves refined communication. He is also the patron of those who have been prophesized, the anointed ones, the saviors. King Arthur as the Once and Future King would fall under his supervision.

He can also help improve your own ability at discrimination, that is to say, developing your capacity to recognize the character and value of each individual and whether they are a right fit for you or not. But, as a side aspect of this, he can also help you develop the power of night sight, the ability to distinguish demons, angels, faerie folk and all sorts of otherworldly beings. Furthermore, he can teach you how to increase your ability to communicate with the otherworldly realms and beings.

And know that his knightly persona is not an affectation. He will defend you if those who are not meant for you happen to try to force their company upon you. If they are too pushy, he will get a bit *bent out of shape* and respond accordingly.

ತ್ತ ತ್ತ ತ್ತ

Arsenic, Lady Arnalas (pronounced air – nay - lace)
Lady of Subtlety
Chant: "With pressure small I will create a change that is so very great."

Arsenic is probably best known for its use in murder mysteries. It has also been used to treat wood and other products as a pesticide, herbicide, and insecticide. It further has found use in car batteries, particularly when alloyed with Copper and Lead and in semiconductor electronic devices in trace amounts to alter electrical or optical properties. Arsenic, in very trace amounts, actually has some positive value in the lives of some animals. However, in any larger doses it takes on its noted toxicity.

Lady Arnalas is the Mistress/Master of Subtlety. She can teach you how very small actions and movements can have very large effects. In fact, she will show you how subtlety is much superior to grandiose movements and gestures, how great things can be achieved with barely discernable force. When she is done, your movements will be efficient and it will seem as though you need do little more than wave your hand to get whatever you desire.

In the past, Arsenic was hard to detect and was known as the poison of Kings and the King of poisons. Therefore, Lady Arnalas can also help you if you have need of being sneaky. We believe honesty is the best policy, generally, but there are times when one must work unseen and move through the world without being recognized for the magical being one truly is. This spirit can help greatly with that.

Lady Arnalas usually appears wearing somewhat shiny but slightly subdued grey, sometimes with yellow or black highlights or tones, and at times with a greenish aura. There may be a slight garlic smell to the atmosphere if you have not filled the room with incense that would cover it up. She has a kinship to Phosphorus' Lord Salfarsey and, as we said, she sometimes allies herself with the Lead and Copper spirits. You, of course, must be careful with her, she needs to be confined to the magic triangle and have no chalice in your circle for she tends to contaminate water, wine, etc.

Because of her association with Copper and Lead, you can have these in the triangle to attract her, but also Iron Pyrite, or fool's gold, can be used as well. Also, real Gold can be used. In fact, the name Arsenic comes from a Persian word that means *gold colored*. She has a love of the peoples, cultures and magics of China, Chile, Peru, and Morocco.

Remember, this elemental spirit is a bit of a demon, and generally associates with Unseelie Fae and those who are less

90

than fond of mankind. Be cautious in dealing with her and make your instructions and desires very clear. Leave no loopholes.

❧❧❧

Selenium, Lady Moncor (pronounced moan - core)
Lady of Renewal
Chant: "I would my magic better be, to long endure
with quality."

Sometimes confused for Tellurium's spirit, Lady Moncor rules Selenium, which is named after a goddess of the moon, and is the matron of reforming and refining one's magic, to make it better, more potent and of greater quality. She has been known to go around with the Copper spirits from time to time. While she may appear by herself, this is actually a rare occurrence.

She can be distinguished by her tendency to wear dark red outfits with tiny black dots. However, they are seldom in a set pattern like polka dots. She is also noted for wearing a lot of beads for jewelry. She tends to favor her hair close cropped, often in a pixie cut. There is frequently the smell of horseradish in the air, particularly if you have a candle or other fire in the magic triangle. She has a fondness for glass so that can be used in the magic triangle to attract her. She has the great capacity to see in the dark, to discern levels of light and power and thus she can help you move, function upon and see the beings of the etheric planes and the subtle planes of existence, as well as discern who is truly spiritual and powerful in the world and the degree of that power.

Selenium has beneficial aspects as a supplement; however, like most things, too much can be harmful. Thus Lady Moncor can assist you with your magic, but even more so with overcoming the tendency to burn yours'elf out, to overdo yours'elf. She will teach you to moderate your energy so it will endure rather than being easily exhausted.

Lady Moncor has a love of Sweden and its culture and peoples. She also has residences in Belgium, Germany, Japan, Russia, the United States and China. Anything from these countries and cultures will be attractive to her. And further, she has, like Arsenic's Lady Arnalas, an association with iron pyrite so that can be used to attract her. She has a deep relationship with rivers and the oceans and goes to them frequently. She also has a working relationship with Lord Sulfur, although more as a replacement. While he might be seen as the Warden of those who have been karmically imprisoned, she is more like the reformer who runs a halfway house for those newly released and learning to use their magic again. She helps them to use it wisely and with moderation.

She will also help you balance yours'elf and your magic. She will tone you down and extend your abilities whether you are too dark or too light in your workings. She doesn't demand that you become a saint, only that your magic be productive for yours'elf as well as others, so you don't fall under Lord Sulfur's power. With her help you will find that you will be able to express yours'elf more truly and with a greater range of ability without accumulating restrictive karma.

Selenium is used in brasses, in association with Bismuth, as a replacement for the more toxic Lead. Thus, once again, this is a spirit that helps one find and choose the light, to make one's witchcraft of a positive and productive variety. Selenium also makes steel more machinable, thus this spirit takes those who are hard and makes them more adaptable. Lady Moncor has an ancient link to the god form of Vulcan (Roman) or Hephaestus (Greek), the lord of the forge and fire and thus she further has a link to Dwarves and all the blacksmith gods such as Wayland the Smith. The scene in the Lord of the Rings where the sword of Aragorn's ancestors is re-forged by the elven smiths, would be under the auspices of this spirit. Thus this elemental can also help you improve the power and quality of your magical implements, giving them greater ability to achieve what you desire.

తతతత

Rubidium, Lord Saråcor
(pronounced sayr – rah - core)
Lord of the Movable Feast
Chant: "I summon thee to make it clear the joy of life
is naught to fear."

Lord Saråcor tends to appear as a slightly overweight gentleman, dressed in silvery white with a dark red, some say purplish, aura, who fidgets or dances constantly, waves his hands around as he talks and gives one the impression that he's going through a mid-life crisis. In fact, he is truly old, older than the Universe and he is just beginning to stabilize and still has his transformative radioactive aspects although these are fading from him. He could be put in with the radioactive spirits but like the radioactive gases, he is moving away from that phase in his life and is mostly, but not entirely, done with it.

There is something soft about him and he is very adaptable as a spirit. He loves to be on the move and he is the Lord, Host and Master of Ceremonies of the Moveable Feast, the never-ending party that goes on forever and travels from place to place. He celebrates life and is the patron of all those who love to party. If you have need of someone to help you with the celebratory aspects of your magic, this elemental can be of great service to you. Remember, the Catholic Mass is called a celebration. Let your magic ceremonies be a celebration as well, for it is ecstasy that is the power and fuel of magic and which connects us to the Divine (see *Shamanism: Archaic Techniques of Ecstasy* by Mircea Elidae).

In spite of the fact that he appears to be slightly overweight, he is none-the-less a very attractive being (think of the *most interesting man in the world*), who draws others to him easily, particularly those who are in need of guidance and direction. He finds many willing supporters and helpers. However, he doesn't care for the compound water, and will react violently if you offer him some. Well, what partiers want to be offered water? He has an easy association and often partners with the spirits of Mercury, Gold, Iron, Caesium, Sodium, and Potassium. However, he doesn't care for the Lithium spirits, even though or perhaps because they are related. Lady Dongur is simply too serious for him.

Rubidium can be found naturally in the minerals leucite, pollucite, zinnwaldite and, of course, carnallite. We knew the word carnal would be found concerning this spirit somewhere. He has a tendency to resist those who wish to bind him to a stable relationship, since he always likes to be on the move, however, if you can reach a compact with him, he will attend to it as he parties along. It's rather like having an interview with someone as sHe walks from hir office to hir car. However, if you reach an accord with him and you set a deadline for fulfillment of your will you can be sure he will be extremely precise in completing his part of the bargain.

He has a fondness for Manitoba region in Canada and has an old estate on the Italian island of Elba, where Napoleon was temporarily held captive. However, like this spirit, perhaps with his inspiration and aid, Napoleon could not be imprisoned there for long and escaped back to France.

He further has association with the mineral Lepidolite so that can be used to attract him, as can a party atmosphere. Lepidolite is said to be able to connect one to the Buddhic plane and has high kinesthetic influence (see *Michael's Gemstone Dictionary*).

ॐॐॐ

Strontium, Lady Cordicor
(pronounced core - dye - core)
Lady Sweet
Chant: "I would the bitterness be gone, let life be sweet
from this day on."

Lady Cordicor, also known as Lady Sweet, can be evoked to make your life a bit sweeter. She is the frosting on the cake, the sugar in your coffee. She can also help you learn to sweeten the deal when bargaining so that what you offer is irresistible and others not only feel good about the deal but also are eager to do your bidding.

This elemental will appear wearing grayish silver raiment that will turn a yellowish gold color the longer she stays. She is extremely active and very willing to make a compact if summoned, or we might say invited (much more polite), by the magician. And you will probably find that she looks, acts and is really quite sweet. And it is said she has a bright smile and wonderful teeth. She can sometimes be found living in the same neighborhoods as Calcium and Barium and comes from the same social circles. The mineral crystal Celestine can be used in the magic triangle to attract her. It helps one adjust to being and functioning on the higher planes of existence where life can be much sweeter.

She is also very fond of Scotland, the Scottish people, cultures and their ancient magic and thus associates with both the Seelie and Unseelie Fae and the Sith of that country, as well as the Pixies or Pict-Sidhe. Anything Scottish is attractive to her, so bagpipe music, kilts or anything that reminds her of those folk will be found greatly pleasing. You can offer her Haggis if you wish, or even better Scottish sweet biscuits or shortbread.

Lady Sweet is further noted for being an entertainer. She may communicate more by pictures and symbols than by words. Look for signs after you have enacted the magic. She is

95

most likely to answer you through dreams after having implanted her messages subliminally.

Additionally, being an extrovert she is unlikely to appear by herself. Both water and Oxygen, particularly if they are together, get her aroused to the point where she is likely to disappear as though a firecracker just went off. You will see a red flash and then she will be gone. So if you want her to stay for a time, don't have water in the magic triangle. On the other hand, if for some reason you wish to get rid of her, although we can't see why you would (can life be too sweet?), tossing water at her will do the trick. If you wish her to remain for a while, have a chalice or bowl of mineral oil in the triangle. She likes things to go easily.

She can also be found at times in Münsterland, a region in Germany, so we expect you could say she has a fondness for the Munsters, that is to say weird, kooky, creepy but sweet people. She further takes delight in the peoples, magic and cultures of China, Spain, Mexico, Turkey, Argentina, and Iran.

Yttrium, Lord Vandcor (pronounced veined - core)
Lord Laser

Chant: "Cut clean away what hinders me, clear the way and set me free."

Lord Vandcor is also known as Lord Laser since he has province over the Light Saber, the magical sword, athame and the witch's curved Boline. All things concerning magical blades are known to him. He can tell you of Excalibur and other enchanted swords of legend and myth (of which there are so many that it would take more than a page to list them all). Additionally, he can inform you how to make and particularly enchant such magical tools for yours'elf or others.

Lord Laser will never come alone. He will always be with a squad, company, group, or fellowship of some sort. He particularly tends to hang out with the rare earth minerals. He likewise has been known to associate with the deadly Uranium spirits. He has an ancient kinship with Sweden and the Swedish peoples and their magic and culture. And the mineral apatite can be helpful in calling him as well as garnets, particularly synthetic garnets. Remember, garnets are often referred to as the warrior stone. Further, cubic zirconia (not to be confused with zircon, although sympathetic magic would allow the substitution if necessary) is another stone that appeals to him.

He will appear wearing silvery robes, looking rather like a jet-eye knight. He often has a reddish hue about him. He is very martial in appearance and attitude, so correspondences for the planet Mars may be used in association with him. But he also has a relationship with the Moon, so moon magic and correspondences may be appropriate as well. He is a dangerous spirit by his very nature and you will want to confine him to the magic triangle. He is noted for liking to stab his enemies in the lungs. While he may not wish to be contained, he has a natural relationship with the triangle, so he will actually feel very comfortable within it. It will be like his little castle.

Note that he often appears wearing a beard. If he appears shaved, it means he's prepared for battle and is more dangerous and greater caution advised. However, Oxygen's Lady Emper has a calming effect upon him, so you may wish to call her first.

In addition to swords, knives, etc. he also has power over needles, scalpels and every other sort of blade and cutting tool, which would include the plough, the sewing machine, drills and everything else that functions by cutting, slashing or piercing. Therefore, you will most likely find that he appears with piercings in his ear, nose or other parts of his body, as well as being tattooed, since tattoos also fall under his dominion. His body is covered with them.

ॐॐॐ

Zirconium, Lady Fadymcor
(pronounced fay - dim - core)
Lady of Wards

Chant: "Seal my circle, strong and safe, no ill can come within this place."

Lady Fadymcor will usually manifest wearing a greyish white gown with a gold tinge to it, covered with various sigils. She is the Master/Mistress/Mastress of magical wards, seals and amulets used for protection and can inform the interested conjuror concerning anything relative to the power of psychic s'elf defense and the protection of one's magical circle and personal being both astral and physical. She is sometimes confused for the elemental of Hafnium, and sometimes, although less so, with Titanium. She is incredibly strong in terms of taking the heat and has the ability to defuse magic that has been cast against her, thus she can teach you how to make your magic circle strong and enduring and enhance your ability to turn away negative projections while allowing helpful and positive energies to be absorbed or passed through. She is extremely adaptable as a spirit and can teach you a sort of psychic Aikido enabling you to adapt to various energies and to redirect them to your advantage.

Like many other spirits, she is an extrovert and believes in the concept of *safety in numbers*. She will therefore urge you to form or join a coven, vortex, lodge or other magical frasority (fraternity/sorority) or to be part of an elven family, band, tribe, circle to increase your powers and your protective aura.

While she has a love of the Earth as a whole, she has an especial fondness for the peoples, cultures and magical traditions of Australia, Brazil, India, Russia, South Africa and the United States. She also has a love of Sri Lanka. Any of these traditions or a combination thereof will serve to attract her to your triangle.

Zircon can be used to entice her as well, and she is most likely to appear wearing one or more of this jewel. Zircon has a wide variety of vibrational powers although the red, brown and blue varieties would probably serve best when summoning her. She also has a strong association with the moon and moon magic and so moon correspondences are appropriate. Rutile quartz can also be used to make the magic triangle pleasing to her. Beach sand can likewise be used in the magic circle since she loves to hang out at the beach. She further has a strong association with the Tin spirits.

As well as teach you how to protect your magic circle from attack, she can further help you make it so that no one can read your mind or know what you are doing using remote viewing or some other astral or psychic technique. Thus, you can protect your circle and magic from spying in addition to intrusion and manipulation. And, if you desire she can teach you how to send intrusive energy back to its source with powerful accuracy. Like Lord Laser, she is expert with swords and knives.

Alas, while she is a great instructor in *defense against the dark arts*, she can be abrasive at times, particularly if you are resistant to learning what you called her to teach. Still, worry not, she will sand you down and hone your magic until you are an expert in using wards, seals, amulets and others forms of magical protection and defense. She will make you stronger, help you feel more secure, and restore your confidence. You just have to tolerate her sometimes rough and other times cutting manner.

> ## Niobium, Lady Unos (pronounced you - nosce)
> Lady Super
> Chant: "Make me greater, better still to best fulfill
> my magic will."

Niobium, sometimes referred to as columbium, is named after Niobe, a Greek mythological character, who is the daughter of Tantalus who was also called the King of Phrygia. Note that elves and pixies and various otherkin are often depicted as wearing what is known as the Phrygian cap.

Lady Unos, commonly called Lady Super or Super Lady, is the matron of all those with super powers. She rules super-heroes of all kinds, and can help you develop your own powers and abilities so they will be truly super, which is to say, superior. However, since the Greek gods punished Niobe for her hubris, we are reminded that modesty is a virtue in the great and in the wee folk.

Naturally, all elfin folk, by the virtue of our magical being, have, in a sense, super powers that Lady Unos can help us increase, refine and develop until they are truly amazing. Whatever your power, she can help you make it stronger. This spirit will usually appear wearing shiny soft gray with a bluish hue, and sometimes there may be a colorful rainbow like aura around her. She will seem to be small and compact, that is to say, she will be small in stature but physically strong and quite attractive with a magnetic personality. She is very adaptable and will transform her lessons and instructions according to your needs and level of development.

She has a natural relationship with Tantalum, which is named after Tantalus, and they bear a family resemblance. She has a love of Brazil and its peoples and cultures, as well as a love of the U.S.A. Her alternate name columbium comes not as one may think from Columbia, the South American country, but from the poetic name for the United States. (Note that

Washington, D. C. the capital of the U.S. means Washington, District of Columbia.) She further has a link to and love of the French Canadian peoples. She also has an ongoing friendship with the Iron spirits and the Nickel and Cobalt elementals. And she makes everyone with whom she associates feel stronger and better. She is a very encouraging being whose help will turn you into a high-powered spirit. She can further assist you to see the world more clearly and in this way use your powers more effectively.

The mineral tantalite, which bears the property of increasing the energy and passion in individuals, may be used to attract her. And you will find her to be a very friendly spirit who poses little danger to you, you need not have your wards strengthened against her, nor do you need to call her within a protective triangle of manifestation unless you so desire. She can be trusted. She is magically hypoallergenic.

ॐ ॐ ॐ

Molybdenum, Lady Setynmacor
(pronounced see – tin – may - core)
Lady of Liaison
Chant: "To spirits that I do not know, with greetings fair
I would now go."

Lady Setynmacor is sometimes confused with the elemental of Lead, but this is in part because in ancient times she would disguise herself as that spirit. She will not appear alone and most likely will come with the Oxygen spirits, if not others. She tends to wear silver with a grayish cast to it. By herself, she doesn't have a great love of water but if she comes with the Oxygen spirits and, as we say, this is very likely, having water in your triangle will be fine with her. In fact, if you wish to ensure that the Oxygen spirits attend her, this is a good way to do it. On the other hand, if you wish

101

to speak with her alone, having carbon and linseed oil in the magic triangle is effective in producing that result.

She is particularly noted, however, for acting as an assistant to the Nitrogen elemental. She is liaison for that spirit and makes most of what it does possible by interacting with plants and other spirit beings. So if you have need of someone to introduce you to others that you would like to know but with whom you do not have a direct connection, she can be of great assistance. She can also help you get ready to function in these new social circles, helping you with a sort of Pygmalion transformation that will make these new folks comfortable around you and you them, so you will easily fit in, if that is what you desire.

Lady Setynmacor is a catalyst of social relationship and interaction. If you don't know what spirits you need to know to advance in your magic, she would be happy to inform you. This is particularly true of all higher spirits. Demons and the lower creatures are easy enough to find and they are eager to meet you and take advantage of you, but higher spirits can be inaccessible at times and she can help overcome this disadvantage. In many ways, she is like the gatekeeper to the higher realms of manifestation. As you might expect, she has a winning smile and loves to display her beautiful teeth.

This elemental also has a natural link to the square and so magic squares are particularly beloved by her and she can teach you much about their use and powers. She is a mage in regards to written correspondence, therefore resumes, curriculum vitae, application forms, letters of introduction, letters of recommendation and so forth, all fall under her purview. She can teach you how to create them so they truly act as a social lubricant to your success.

The mineral wulfenite can be used in the magic triangle to draw her. This has the power to increase one's ability to speak with the Devas, so it is doubly efficacious. And she has a particular fondness for the people and magical traditions of China, the United States, Chile, Peru and Mexico, although she can be found dwelling in other areas as well, since she is a sort of social butterfly, such as Norway, Colorado, British Columbia and Utah.

ॐॐॐ

Ruthenium, Lord Hawart (pronounced hay – ware-t)
Lord of the Covenant
Chant: "Let us make our compact be strong, to last,
endure forever on."

Ruthenium's Lord Hawart is the master of the Covenant, pact and all magical agreements. By extension, he has province over contract law and contract lawyers of all sorts, but he particularly oversees magical agreements. Thus, if you pray to some deity and promise you will do such and such if the god/goddess saves your loved one from so and so, this is a magical contract that he will oversee to its fulfillment. He can inform you of anything you wish to know about magical compacts and covenants.

If you think you have gotten the short end of the stick in a compact, and especially if you feel you have been cheated or the spirit hasn't done its part, you can appeal to Lord Hawart to look over the contract. However, he will go strictly by the law and he cannot be corrupted or bribed to do otherwise. Ruthenium is part of the Platinum group on the periodic table, a group that tends to be a bit snooty and do not care much for associating with others of what they perceive to be lower class spirits. They hold themselves in high regard. Although, Lord Hawart is sometimes known to hang out with the Copper and Nickel spirits, who are a bit more middle class.

Named after Ruthenia, a land founded by the Swedish and Finnish Vikings, and the source of the word Rus from which the name Russia comes, this spirit has a distinct love of Viking culture.

103

Lord Hawart will manifest wearing silvery white and appears to be a very strong and robust individual, again think of a very large, muscular Viking. He will often be wearing jewelry made of Platinum or Palladium. He is aware of being a very rare and special being. In other words, he thinks like a lawyer.

He is especially fond of the Ural Mountains and of North and South America, also Ontario, Canada, and South Africa. He is long noted for having a love of Pre-Columbian art, artifacts and culture. He also has association with the asteroid Vesta named after the virgin goddess of home and hearth from Roman mythology, but again that merely speaks to his incorruptibility. He is a rare lawyer, indeed. He is also a very enduring spirit, he will not give up until he is done and his influence will similarly tend to strengthen the endurance of the conjuror who makes a compact with him.

However, once again note that he goes by the spirit and letter of the law. If you have signed a contract with a demon or some other denizen of the underworld, you need to fulfill your end of the bargain. He will not help you cheat the devil. On the other hand, if you are compelled into association with such creatures, he can assist you in interacting with them without losing your soul.

If you do enlist his service, you won't be called on to sign the contract in blood, rather, he will most likely expect you to sign with the impress of your fingerprint or thumbprint. You can use ink if you wish, but you don't need to do so because he can read latent fingerprints and merely pressing your finger to the paper will be enough. The oils on your hand will do the rest.

Rhodium, Lord Sascor (pronounced sayce - core)

Lord of Investment

Chant: "Guide me so I may invest and get from it
the very best."

This is a very rare, valuable and elite spirit. He is the patron of investment of all sorts, thus he rules Wall Street, hedge fund managers, stockbrokers and bankers. Anything you need to know about investment, he can tell you. He can also aid you in investing so your investments are successful. This extends to investing your magical life energy. If you follow his advice, things will start to happen. He knows how to get things moving. He has the magical ability to convert small investments of magic into huge outcomes.

Lord Sascor manifests wearing a silvery white suit often giving off a rosy aura. The name Rhodium comes from the Greek word for rose (the rose perhaps being a good symbol for the stock market, a sweet smelling and beautiful flower with sharp thorns). He gives one the impression of being a very hard and serious individual and he doesn't much care for involving himself with others outside his social circle. He has a sense of being of noble birth, coming as he does from a very wealthy family. He is well aware of being an elite spirit. However, he keeps the interests of those with whom he associates in the forefront of his mind. If you come to an agreement with him, you can trust him. He is nearly incorruptible, unlike worldly bankers, and very protective of those under his care.

He also has a tremendous ability to understand and reflect the market and to invest and move money around accordingly. He is a wizard of Wall Street and all such investment institutions. But, if you don't have the wherewithal to invest in the stock market, he can show you how you can use what you do have wisely, how to buy and sell and make a profit, how to increase your wealth at whatever level you are currently upon. And unlike so many others, he is a totally reliable spirit. He doesn't panic during market fluctuations and in fact has been known to make a fortune when others were losing their shirts. He is a rare being, indeed.

As you might expect he is noted for associating with the elementals of palladium, silver, platinum, and gold and any of these metals can be used to attract him and he may very well appear wearing jewelry made from these precious metals. He has a fondness for South Africa, Russia, particularly the Ural Mountains, and the Ontario region of Canada.

While he may seem snobbish, and we don't deny this, he is not a dangerous spirit generally. He is unlikely to harm you in any way. Gaining his trust, however, is another matter. At the same time, like most people in investment, he wants to know what you have to invest. Come to him with what money or energy you wish to invest or increase and he will barely be able to keep himself from advising you. He's clever and he knows it and likes to demonstrate his powers. You may think that a compact with him will cost you a lot, but the truth is, he's too rich to care. He does it for the love of doing it.

నానానా

Palladium, Lady Hasarcor
(pronounced hay - sayr - core)
Lady of Lucky Accidents
Chant: "Watch over me and let me find the silver lining that is mine."

Manifesting in shiny silver white, sometimes with a rust colored aura, Lady Hasarcor is the elemental of lucky accidents. By this we mean that she can help you find the silver lining in all the dark clouds of your life, as well as promote synchronicities and coincidences that will have a very positive affect upon your life and magic. You may have suffered what has felt like a setback in your life, but this spirit can show you how the balancing aspect of that event will prove beneficial to you. She can teach you how to turn the negative into the positive, the destructive into the beneficial, and a loss into a gain. She is a healing spirit and can help you repair any damage to your body, your circumstances, or your spirit.

Lady Hasarcor has an association with the asteroid Pallas, the Greek mythical character Pallas that the asteroid is named after and by extension the Greek Goddess Pallas Athena. Any and all correspondences associated with these can be used in her evocation.

Palladium is part of the platinum group, composed of Palladium, Platinum, Rhodium, Ruthenium, Iridium and Osmium, that clique of rich and snobby elementals that tend to keep to themselves. However, that doesn't mean you, amazing sorceress, conjuror or magician that you are, can't enlist their aid in your magic. You simply need to know how to enchant them. And being an elf or fae of some sort, you are an enchanter by your very nature. Lady Hasarcor is the most accessible of this group, the most sympathetic, empathetic, and lightest of spirit. She is of noble birth but takes the idea of Noblesse Oblige, the responsibility of the powerful and wealthy to help those less fortunate, very seriously.

She can be commonly seen in South Africa, Montana in the United States, the Ontario region of Canada, South America and in Russia. She has a love of all these peoples and their cultures and magical traditions. She is also known to frequent Australia and Ethiopia.

She is very big on recycling. So if you have magical implements or raiment that you have made from other things, or that you purchased second hand, she will very much approve. Unlike so many elementals, she has little to do with the Oxygen spirits. She will listen to them, but is generally unmoved by them. On the other hand, the Hydrogen spirits tend to love her and follow her around, getting very excited just by being near her. They are big fans of hers.

Lady Hasarcor is noted for loving flute music so if you play the flute or put on flute music for your evocation, this can be very helpful. She will most likely manifest wearing jewelry of her own making, as well as, having gold or silver colored teeth. You may also have flowers in the magic triangle as an offering; however, she is known to have a dislike of the water hyacinth, a plant that is native to the Amazon. For payment, it would be well if you did your best to help others who have suffered due to no fault of their own.

ॐॐॐ

Silver, Lady Arvyn (pronounced air - vin)
Lady of the Shining Ones
Chant: "Oh, bright spirits hear my call, make me greater overall."

Silver is the element most commonly associated with the elven, particular the high elven (and, let's face it, many elves are high; we are high by our very natures). Tolkien's Mithril or Elven Silver, however, was probably an alloy of Silver and Rhodium. Lady Arvyn, the elemental of Silver, can assist one in gaining access, guidance and aid from the Shining Ones, those evolved and advanced elven beings that are most like the elves in Tolkien's books. They are equivalent to what most people think of as guardian angels. These beings can help you with any aspirations you have toward higher magical and spiritual advancement. They can also help you become more successful in the world, on the proviso that your material advancement also furthers your spiritual evolution. (See our book *The Shining Ones: The Elfin Spirits That Guide You According to Your Birth Date and The Evolutionary Lessons They Offer.*)

They can also help you empower your magic, clear your emotional centers, see and remember more clearly, as well as,

108

glimpse into the astral and etheric realms of being, and help you remove any parasites that have attached themselves to your physical or spiritual being. They can further aid you to become much more adaptable, a magical power that is often underrated. With the help of these spirits you will become more powerful, more efficient, and the time between the act of magic and the results will be greatly shortened. These beings have an extensive memory as their lives are much longer than ours and their memory retention is profound. Mere contact and association with these spirits will uplift your own spiritual status and development. The saying *when the student is ready, the master comes*, is relevant here.

Like the element itself, Lady Arvyn will appear wearing shiny silver colored robes with a slight whitish hue. She has also been known to wear yellow or orange scarves. If she stays for more than a few minutes her robes will tend to darken, creating deeper and richer shadows in the folds. While she may come alone, she is known for hanging out with the Copper, Gold, Lead, and Zinc spirits. She has a fondness of stained glass, so if you have stained glass you can use that to attract her, or if you can somehow call her in some church or other setting with stained glass, that can be effective. Unlike some faerie tales that say elves cannot enter churches, the truth is, we simply find most church services boring. Churches themselves, particularly old style gothic churches, are very attractive to us.

This elemental is, of course, known and loved worldwide. For her own part, she does have an especial fondness for Peru, Bolivia, Mexico, China, Australia, Chile, Poland, Tajikistan and Serbia and their peoples, cultures and magical traditions. And she has long had an association with the Navajo peoples of the Southwestern U.S. She is also known to have a long relationship with the Ancient Judaic people and the people of Babylon. Naturally, objects made of Silver, either solid or plated, can be used to attract her.

෨෨෨

> ## Cadmium, Lady Nacor (pronounced nay - cor)
> Lady of Appearance
> Chant: "Better looking I would be, from now until
> Eternity."

*U*sually, Lady Nacor will manifest wearing bluish silver garments often with a red, yellow or orange hue. She is beautiful and almost everyone gasps when coming upon her. It's like unexpectedly encountering the most gorgeous, glamorous, sexy and radiant movie star on the street. She rules all things that concern appearance. Thus she has dominion over plastic surgeons, fashion designers, make-up artists and anyone and everything that affects the way you look, including those who promote exercise as a means of improving your appearance. If you have a desire to transform how you look, to alter and improve your style, then this is the spirit to summon.

She has long had a friendship with the Zinc spirits. She also has an association with sea sprites. She has a love of the peoples, magic and cultures of China, South Korea, Japan and, curiously, Siberia and thus has a protective feeling toward Eskimos and various tribal people of that area including the Sami people of what was once called Lapland, the reindeer people. Thus she also, by extension, has a relationship with Santa Claus and his elves.

However, Cadmium is poisonous, so remember that one can mistake appearance for reality, can judge books, and people, by their covers, can fall into the narcissistic tendency to believe one is greater than one is merely because others find one attractive. Remember, beauty of form is important, it adds to one's life, but it is not the most important thing. Don't sacrifice, or forget, substance in favor of appearance. Just as fancy words are not as potent as those that are truly heart felt, so fancy clothes are not as significant as the integrity and character of the person wearing them. Yet, that doesn't mean

110

one should neglect one's looks just because it is less important than, say, one's health. In fact, while this spirit will help you improve your visual presentation she will also advise you how to make this change of lasting value to you. She is a great proponent of regular exercise.

Cadmium has long been used as a pigment in paints, and even now when Cadmium is being phased out of use because of its toxicity, there are still paints called Cadmium Red or Cadmium Yellow, although now they often have the appendage *hue* attached to let one know that the actual Cadmium is gone but the richness of color remains. So, too, this spirit will bring color and richness to your life and appearance, but keep her within the magical triangle for safety. If you use sigils in her evocation, paint them with red, orange or yellow.

However, while she can be dangerous, she also has her protective side. So if you do develop a relationship with her, she will make sure that your visual persona is potent and successful. She will demonstrate how to make it so you radiate strength and don't give off signals of being a potential victim. In fact, she can inform you how to dress or present yours'elf in different settings and social circles in order to stand out or fit in as best serves your spiritual and magical purpose, how to dress for success or for a party. She can also demonstrate how to make your physical form arousing or soothing as needed. In that way, she has the great power to heal difficulties between individuals that have resulted from rash words and actions. Since nearly everyone finds her attractive, they are eager to listen to her.

Indium, Lord Newor (pronounced knee - wore)
Lord of the Sacred Vow

Chant: "I dedicate my life henceforth, in friendship, love and all of worth."

Indium is a rare element that has high fusibility and is often used in solder. Lord Newor thus rules marriage but even more than that he oversees the Sacred Vow, the dedication of one to one's magic, vision or some great goal. In terms of marriage, his influence is toward a true, deep and enduring relationship. By extension, he also has influence over elven groups, bands, vortexes and covens that make lasting vows of relationship and association. He will help you turn a family of choice into an enduring family. Don't call him and make these vows if you don't intend to keep them. The consequences can be painful, and more so disempowering for those who abandon their magical vow.

Lord Newor will manifest in lustrous silvery white, often with a bluish indigo aura. In fact, the name Indium comes from this indigo coloring. Thus he has affection and feels responsibility for the Indigo Children. He appears soft and is, in fact, quite understanding and adaptable. He knows that life and relationships are not easy, and that life circumstances may change and he is more than happy to inform you of this before you make a vow and almost certainly will ask you several times if you really want and are determined to fulfill the Sacred Oath. However, if you proceed, know that he will encourage you strongly to remain true to your vow, or face the possibly of a very large karmic penalty fee for early withdrawal if you don't. And he is not beyond screaming at you if you waver, if only as a warning that you are doing so. He will also most likely make you write out your vow on paper and sign it, to underline the seriousness of what you are undertaking.

This elemental has a love of glass, so a glass chalice can be used in your magic circle or triangle. However, this spirit has no interest in water, nor does it care for interacting with Boron, Silicon, or Carbon, and doesn't much care for the Hydrogen spirits either. He finds them dull and uninteresting. He has an ongoing relationship with the Zinc, Tin, and Copper spirits. He also has a fondness for China, Canada, Japan, South Korea and Bolivia and their peoples and magical traditions.

Indium has low toxicity, so this spirit is not dangerous to you, even if you don't fulfill your promise he'll simply leave you to the karmic result. As we said, the most he's likely to do is to cry out urging you to continue on, but if you choose not to do so, that's your business. He's done his part and the rest is between you and the Divine Magic.

He has an ancient relationship with the Saxons, and considers vow taking just as seriously as those folks did, which is pretty seriously. About as serious as a heart attack. If you don't really want to fulfill the vow you intend to make, or if you think that you only mean it for a short time, don't waste this spirit's energy and time. If you say your marriage vows with *until death do us part or we get tired of each other or find someone we like better*, you don't need this spirit to seal the deal. Alternately, if someone has made a sacred vow to you and has failed to live up to hir end of the bargain, he'll be glad to look into it.

Wulfenite, which helps one connect to the Deva spirits, can be used in the magical triangle to attract him, as can apophyllite, autunite, vesuvianite, zircon, and rutile.

> ### Tin, Lord Naf (pronounced nafe)
> Lord of the Magic Numbers
> Chant: "The logic of the world is math, teach me
> of this magic craft."

While this tin soldier has province over security guards and rent-a-cops, his true power has to do with the magic of numbers, the power of mathematics, algebra, and calculus to understand the Universe, and the way the world functions (see the TV show *Numb3rs*.). Thus this elemental is related to that great and powerful magus of Ancient Ionian Greece, Pythagoras.

Now, these elves admit that we are rather lacking in this department. We've had little relationship with this spirit, and have small understanding of math as a magical power, and are unlikely to pursue this study further in this particular lifetime. However, there are numerous magicians who are deeply into this magic, and we admit a bit of a deficit in our magical education in having done nearly everything we can to avoid these magical lessons. (We saw a cartoon today of a librarian re-shelving books asking, "Who keeps putting the math books in the horror section?" That would be us.) However, if you are interested in using mathematics as a magical power, then this is surely the spirit to conjure. Unlike Lady Setynmacor, who can teach you of magic squares in particular, this elemental knows everything about the magic of numbers overall, and has been called doubly magic so great is his power.

The mineral cassiterite, which is another warrior stone, can be used in summoning this spirit particularly if you are doing so to evoke its protective qualities and would like to hire a spirit to guard you and yours. Note, however, that you will have to pay this spirit, or spirits, regularly and as soon as you stop they will quit and go their way. This is simply protection for hire and these are not spirits that act out of patriotism or any idealistic intent.

This elemental will generally manifest wearing silver white robes, often with a grayish tint. He is noted for carrying a long thin stick, which he uses to point to things when he teaches and is known for slamming it down on tables or desks or blackboards producing a loud cracking sound if you displease him or your attention wanders. In other words, he's an old-fashioned math teacher. However, his bark is worse than his bite and you don't really need to be afraid of him. He does, however, like to eat, so if you have a full place setting in the magical triangle with offerings of food this will be very pleasing to him. Also, he is very partial to pewter. So using pewter cups, plates, etc. will entice him to your triangle. Although, if you have cans of foods, tin not aluminum, this will also serve to draw him to your magic.

He further has a partiality to diamonds and all crystals of a cubic form, such as diamonds, garnets, fluorites, pyrites and spinels. He also has a long association with granite. All these can be placed in the magic triangle and are more appropriate for summoning him for his mathematical prowess, except for granite, which is better used for his protective aspects. He has a fondness for rivers, so if you do an outdoor evocation, doing so close to a river will be very conducive, especially if you can do it in the bed of a shallow stream or on an island.

Lord Naf has also been known to love the peoples and magic traditions of China, Indonesia, Peru, Bolivia and Brazil. Although, it has been recently rumored that he might be acquiring new residences in Mongolia and Colombia.

Antimony, Lady Matela
(pronounced may – tee - lah)
Lady of Feminine Wiles

Chant: "I seem frail so you do not see how very powerful I can be."

Lady Matela was first mistaken for the elemental of Lead and whether she deliberately was disguising herself, which is our opinion, or it was an error of those who encountered her, we do not know. She will materialize wearing a shiny gray gown with very elegantly applied makeup. Most likely she will wear kohl eye shadow. In fact, the name of the element is most likely derived from her invention and use of kohl. She often wears six rings on her fingers, generally three on her right hand and three on her left, which are usually of Silver. She is the matron/patron of all those who are *the power behind the throne*. President's wives, king's mistresses, and every person, most likely but not exclusively female, who has really ruled a company or country while bathed in the shadow of obscurity, is under the dominion of this elemental spirit. She is the *woman behind the man*. She is known to work in concert with Lord Naf, although she is generally considered to be even more attractive than he is.

If you have need of influencing such a being, or becoming one, she can help you do so. She will teach you how those with intelligence can dominate and influence those who have position and power. She will show you how to use the traditional feminine wiles of receptivity and beauty to guide those you wish to influence. How one doesn't need to be in power to affect and persuade the powerful. How to entice them, soothe them, or arouse them as needed. How to dampen their anger and how to redirect their passions toward the fulfillment of one's chosen goals.

She has an enduring relationship with the people, cultures and magical traditions of Ancient Egypt, Caldea and China, and in the course of time she gained influence in Ancient Greece and Rome. In modern times, she has found a home in Sweden, also in Myanmar, Canada, Tajikistan and Bolivia. While she is a rather rare being, her associations are wide. Antimony has relations with over 100 different minerals, which tells us this elemental has a wide web of allies.

This spirit is known to be a wily being who is, none-the-less, fairly stable, which is to say she is not easily upset by changing circumstances. She knows how to go with the flow and how to make strong and lasting alliances. She is resistant to attacks from jealous, envious and acidic individuals that sometimes surround her and she can teach you how to ignore and be unfazed by backbiting and acerbic comments. Yet, she appears to be a very soft being. In fact, it is her very softness that is her greatest power. She may appear to be quite sensitive or frail, yielding easily, but she knows what she wants. She has an aspect of explosive power that she rarely reveals unless forced to do so. She much prefers persuasion and allure.

Antimony is poisonous, so use your magic triangle for summoning her. And, it should be noted that those of the dark arts who have called this spirit through history have often used poison as a weapon. This would include Lucrezia Borgia; Livia, the wife of the first emperor of Rome, Augustus; and the fictional character Milady DeWinter of the *Three Musketeers* saga.

చాచాచా

Tellurium, Lord Arthorcor
(pronounced air – thor - core)
Lord of the Golden Pathway
Chant: "I dedicate all I do to bring about my vision true."

Tellurium's Lord Arthorcor is much more comfortable out and about in the Universe than he is on earth. He is a spiritual being, by which we not only mean he is a

117

spirit, but that his primary concerns as a spirit are of an evolutionary and spiritual nature. He is the Lord of the Golden Pathway, sometimes called the Yellow Brick Road, which leads one to the fabled city of myth and legend. He rules all monastic people. Thus he has province over monks, nuns, priests, priestesses, hermits, yogis and all those who have, in as much as possible, renounced the material world and dedicated themselves to spiritual concerns. You don't often find elves among these, but it is not impossible, for there are those elves who retreat from the world. If you wish to dedicate yours'elf to the spiritual path and put aside social and material advancement, or if you have need of contacting, influencing, or dealing with such beings, this is the spirit who can best fulfill your will in that area.

This elemental will materialize as a somewhat stiff looking old gentleman wearing silvery white with a reddish aura giving off the smell of radishes. He may remind you of an elderly librarian of the stuffy sort, or perhaps a monk. He gives off the aura of great stability. You get the feeling you have to prove to him you're worthy of his time. He has a familial relationship with the Oxygen, Sulfur, Selenium and Polonium spirits. He also has a relationship with Gold. However, he is most often known to associate with the Copper and Lead spirits. He is noted for helping other spirits get some spiritual backbone but also to become more spiritually flexible, thus less dogmatic. He constantly urges aspirants to upgrade their spiritual understanding and beliefs, getting rid of whatever is no longer useful to their evolutionary spiritual growth.

If there is a garlic smell, when he comes, instead of the radish aroma, it means he's not in the best of moods and since there is a moderate toxicity to him, manifesting as a scornful opinion of material concerns, it is best to call him within the magic triangle. Also, it's no use asking for anything worldly from him, neither material success nor social advancement; he just doesn't care. His duty, as he sees it, is to help you align to your spiritual destiny and to help you find your place in the

Cosmic scheme. He does, however, have a fondness for crystals, so you can use those in your magic triangle. Also, gold and the vibrations it gives off will be well received. He is known to be fond of ceramics, particularly homemade and colorful pieces, so these can be utilized as well.

He has a fondness for Romania, especially the region of Transylvania known as Alba. Alb is root word in German for elf and Transylvania is the home of the Dracul, the vampire people and the elven dragon people (see Laurence Gardner's *Realm of the Ring Lords* and DeVere's the *Dragon Legacy* for more on this). Thus this elemental has dominion over vampires and dragon folk and can help you deal with any beings who attempt to drain your energy or your resources or attempt to block your spiritual progress or lead you astray. He has also been known to frequent United States, Peru, Japan and Canada and take delight in these peoples and cultures.

࿐࿐࿐

Iodine, Lady Rodfaso (pronounced road – fay - so)
Lady of Healers
Chant: "I heal myself and vibrate, too, the healing that is
good for you."

This elemental will take visible form wearing shiny bluish black robes with a violet purple aura. Lady Rodfaso has an affinity for water, particularly salt water, so you can have this in your magical triangle. She is the Master/Mastress of healing and healers. Thus she has dominion over practitioners of the modern medical arts as well as traditional forms such as Reiki, Qigong, Tai Chi, when it is used as a healing modality, and of chiropractors, doctors, nurses and every other form and being of the healing arts and sciences. Therefore, she also has association with Jesus in his manifestation as a healing mage, also Asclepius, the god of medicine in ancient Greek religion and mythology, and his daughters: Hygieia who represents Hygiene, the goddess of health through cleanliness and sanitation; Iaso, the goddess of recuperation from illness; Aceso, the goddess of the healing process; and Panacea, the goddess of the universal remedy (most likely love or love magic). Also, she has relations with the Roman/Etruscan god Vediovis and all other healing gods and goddesses from every culture and tradition, so numerous it would take more than a page to list them all.

Besides ruling health and healing, she is especially a matron/patron of intellectual development, the proper development of the brain, and of the nurturing and maturation of the individual. Thus she also has influence over diet as an aspect of maintaining health, of healing and of the growth of the individual. Her concern is not only of the health of the body but of the mind and the psyche, therefore she also has power concerning psychologists and psychiatrists.

She further has some power in the banishment of demons, particularly if some small darkness has crept in and you need to exorcise it or you have found an opening in your magical field and need to seal it quickly. Furthermore, she has a very discerning nature and can detect when someone is trying to scam you. So if you are usually a victim or target of such scams, she can help you to spot them.

Because Iodine normally exists as a diatomic molecule, a molecule of two atoms linked together like a rod, and because the symbol of healing is the Rod or Staff of Asclepius, a serpent-entwined rod, Lady Rodfaso also has power over magical rods and staffs, particularly when they are used for healing or teaching. (This rod is sometimes confused with the winged caduceus wand carried by the messenger god Mercury.) If you wish to create and/or charge a magic wand for healing purposes, this is the spirit to summon. The snake, of course, is a symbol of esoteric knowledge.

This spirit has a fondness for the peoples and cultures of Chile, Japan and the state of Oklahoma in the United States.

She also has a relationship with the gas elementals. Moss and fern agates, which have healing powers, can be used to attract her.

෴෴෴

Caesium, Lord Mercor (pronounced mere - cor)
Lord of Time

Chant: "Help me master magic time, and quick fulfill my will's design."

Known as the Lord of Time or sometimes as a Time Lord, Lord Mercor can help you master time, which is to say develop longevity by approaching the speed of light and increasing your ability to move through space efficiently and with decreasing resistance and friction. He can further aid you to shorten, again an aspect of increasing your vibration rate, the time between the execution of your magic and the fulfillment of your will.

Lord Mercor will arrive wearing silvery gold colored clothing, which will tend to darken the longer he stays. He radiates a sky blue aura. He has a rather genial look to him and tends to laugh and smile a lot. Since, Caesium is one of the five metals that have a propensity toward being liquid at close to room temperature, this spirit has close association with the Water/Liquid elementals and can sometimes be utilized for its liquid power. However, know that it is extremely reactive to water and air and will ignite when coming into contact with them, thus Lord Mercor can also be said to have a relationship with the Fire/Illumination elementals. He has the power to easily cross barriers and boundaries, going from one parallel world to another, one plane to another, and can teach you to do the same. However, when you really attain the mastery that this spirit teaches, you will more than likely shift to subtler planes of manifestation and vibration that are closer to the

121

speed of light and will depart the world you are currently in altogether.

Lord Mercor can also inform you about living in the vacuum of space and how to effectively direct and use the power of electricity and light. If you've ever seen or read in movies and books about wands that shoot bolts of lighting, then you are observing his power and work.

While he has close association with Gold and Mercury, he doesn't care much for Cobalt, Iron, Molybdenum, Nickel, Platinum, Tantalum or Tungsten. He has been known to hang out with Antimony, Gallium, Indium and Thorium, however, who share his interest in light, which is to say power, time and transformation. Although generally speaking, other spirits are of the opinion that he is such an introvert that he is, for the most part, incompatible.

Because of his explosive nature in the material world, it is best if one has some mineral oil in a bowl within the magic triangle, otherwise, he is likely to ignite and disappear. If you wish to bind this spirit, stainless steel is conducive to doing so. This is the type of elemental that, like a genii/djinn, could be contained in a hermetically sealed object, again of stainless steel, although, these elves never attempt to bind spirits unless absolutely necessary.

This spirit can most often be found in Manitoba, Canada; Zimbabwe; and Namibia. The minerals lepidolite and petalite are beneficial for conjuring him. Also, he has a fondness for mineral water and that can be used in the magic triangle.

**Barium, Lady Sylwima
(pronounced sill – why - mah)**
Lady of Gravitas
Chant: "Power I will now exude, profound and deep
my every mood."

Lady Sylwima is the elemental to call if you wish others to take you seriously and to see you as an individual with gravitas. Other spirits frequently refer to her as being "deep" and "heavy" and as "an elemental who is truly profound". You don't get the idea that she is depressed or brooding so much as the sense that she takes herself and life very seriously and has no time to waste. Therefore, if you are ready to take life and your magic seriously, then you may wish to summon this spirit. However, know that she will aid you in as much as you demonstrate your dedication and intent to truly developing yours'elf as a magician/conjuror. Witches, alchemists and psychics through the ages have lauded her aid and say that she imparts a glow upon them and their magic that lasts for years.

Her material presence will tend to come in the form of an attractive but quite serious being, wearing dark silvery white clothes with golden shade to them. She is known for having a green aura. She radiates a lot of energy. Her mere presence will make you feel that something profound is taking place and she can teach you how to have the same effect on others.

She will not come alone. She is always surrounded by a coterie of spirits that assist her or wait upon her every word. She sometimes associates with the Aluminum, Zinc, Lead, and Tin elementals. She is not partial to water, nor alcohol really, she just doesn't have time for frivolity. You get the sense that she is incredibly busy and that you are privileged to have a few minutes of her time. Don't waste your opportunity. Simply acknowledge that you are ready to dedicate yours'elf and she will do the rest.

She has been known for making the steel and cast iron spirits, who are generally noted for their strength and seriousness, feel like they were weak until she came along to sort them out. She is known for despising rodents and will tend to kill any that are about. So don't have your pet hamster in the same room where you evoke her.

However, don't mistake her for a being who demands absolute purity. She is not opposed to that, but she is much more concerned that your will be strong, your intent unwavering, and your dedication sincere. She is less bothered about foibles and idiosyncrasies than about the continuing effort to develop yours'elf and your goals. You don't have to be perfect, merely consistent in your efforts.

While you don't want to have water or alcohol in the magic triangle, you can have oil, particularly body oils. She likes to keep her skin protected. You could even have perfume oil, but only a dab. Like Tin's Lord Naf, she has an appreciation for cubic-formed crystalline structure so diamonds, garnets, fluorites, pyrites and spinels can be used in the magic triangle. Also, the mineral witherite, which is said to help one control compulsive behavior, may be used to attract her.

She has had a long association with Bologna, Italy. Also, she is often found in England, Romania, China, India, Morocco, US, Turkey, Iran and Kazakhstan and has a great love of their peoples, traditions and magic. And while she doesn't care for fresh water, she does have a partiality for the sea.

Lanthanum, Lady Naena
(pronounced nay – ear - nah)
Lady of Fame
Chant: "I would be known, oh, this is true, let honor
       and respect accrue."

I f you hunger for fame and recognition, if you desire to be in the spotlight, then Lady Naena is the elemental to summon. She is a publicist supreme and can get you the attention you wish. If you are an artist, a writer, an actor, a musician, a politician or a scholar and you feel you are not getting the recognition you deserve, she can be of great help. What you need to ask yours'elf is: do I really want to have fans and all the attention and responsibility that having them entails?

She is known to hang out with Cerium, in particular, and the other rare earth elementals, which as we've said elsewhere are not actually rare on Earth but perhaps we might consider them somewhat eccentric. She favors wearing silvery white clothing with a greenish tint.

The word Lanthanum comes from the Greek word for *to lie hidden*, and what this elemental does is expose what has been hidden and bring it to the light. Thus this elemental has the power to expose wrongdoing, to make infamous the villainous so their ill deeds are noted and dealt with. This elemental has a penchant for purity so know that if you have done wrong, it will be revealed. So once again, ask yours'elf: do I really wish to be famous?

This spirit has a great affinity for California but has also been known to have affection for the peoples, cultures and magical traditions of Africa and China. This spirit is a promoter of hybrid cars and very much supports those who work toward a more sustainable world and culture.

Lady Naena further has the ability to recognize high-energy people. She knows who is likely to succeed and become famous in any profession. She can also help you distinguish between those who will energize your life and those who will merely drain your energy. She has a tremendous ability to recognize talent when she encounters it and can tell you who is likely to move to the top quickly, who will burn out like a comet, coming suddenly to fame and then disappearing, and those whose influence will endure. She has an especial interest in the motion picture industry, but in fact can speak to you of anyone concerning any field of endeavor. She also has a fondness for photographers of all sorts, and a love of astronomers. She furthermore has some interest in archeology, but this is more of a side study than a main one for her.

This elemental also has some healing abilities, and is particularly devoted to those who suffer from renal disease. You may say that she contributes to research toward dealing with and overcoming kidney problems. She particularly has the power to control and banish excessive phosphate spirits, who can prove to be a problem in this and other areas. However, be warned, this spirit often affects individuals in such a way as they end up with hyperglycemia, so if you do conjure her, don't forget the importance of maintaining a balanced diet; don't let fame go to your head or your waistline.

Zircon, which is noted for attracting fame and fortune, can be used in the magical triangle to attract her, as can Amber, which aids in healing renal disorders.

Cerium, Lady Rongra (pronounced roan - grah)
Lady of Refinement
Chant: "I would to polish what I do, refine my magic bright and true."

Named after the dwarf planet Ceres, which in turn was named for Ceres, the Roman goddess of agriculture, this elemental has influence over both dwarves and farmers. This influence over dwarves is doubly so due to the fact that Cerium is used in jeweler's rouge, which is utilized for polishing gems, and dwarves, of course, are noted for their love of jewels. Thus this spirit also has dominion over jewelers. But her major power is the power of refinement. If you have need of polishing your work, your personality, your magic, or any other aspect of your life, she is the elemental to summon. She will burnish away your rough edges and make you shine. She is extremely adaptable and it doesn't matter what field you are pursuing, she most likely has knowledge concerning it and how to improve your work within it.

She is known for being the head mistress of Lady Rongra's School for the Refinement of Young Enchanters. She is said to have a magnetic personality, and can teach you how to become permanently magnetic, attracting others to you with little or no effort. Of course, she is noted for having helped many theatre and movie personalities with their careers. She can also teach you how to keep up your energy level so you are not worn down, nor have your aura darkened by extensive involvement with the public. She can further teach you, if you are interested, how to use makeup so that you look good under stage lighting, so that you give your best appearance in theatre, movies, photographs or television.

Lady Rongra will appear wearing a quiet silver gown. Sometimes, she will be sporting an orangish red or yellow hat. Her skin is often said to be very clear, and thus she looks a bit like a porcelain doll given life. She is known to smoke (what we are not divulging) and thus is often seen carry a silver-colored lighter and a long black cigarette holder with silver tips. She is, however, affected by the temperature. She prefers heat to cold and is much slower to appear in colder climes. While it may not be apparent to you, she has the tendency to absorb ultraviolet light.

This spirit also likes things to be clean and tidy. So before summoning her, make sure you have straightened and dusted your magic room. This, by the way, will demonstrate your intention to put energy toward the refinement of your character and art and will be much appreciated by her.

She has a love of and a long history with the people and cultures of Sweden and Germany. She is known to have an association with the elementals of Sulfur and Oxygen and through them, as a friend of a friend, with the Aluminum and Iron spirits. She is known to like barley tea, so you can have that as an offering in your magic triangle. She is also known to have a relation to the Beech tree, called the mother of the forest, whose planetary symbol is Saturn, so Saturn correspondences can be used in your magic. The Beech is also associated with writing, so don't think this elemental only deals with public appearances, she can also assist you to refine your writing and communication skills.

Since Cerium has low to moderate toxicity, it is important to conjure her within the magic triangle. And she can have a bit of a temper at times, although she is likely to express it with scathing wit. Some have suggested that she was the overshadowing spirit for Dorothy Parker (please google her).

≈≈≈

Praseodymium, Lord Genlascor (pronounced geen – lace - core)
Lord of Patience
Chant: "Patient I will ever be, my will fulfilled, in time
set free."

Manifesting in silver velvet with a slight golden tinge, this spirit is noted for being very adaptable and, of course, as its power suggests, patient. If he stays for any length of time, you will most likely observe a greenish aura about him. Some have said that he is flaky, which is not to say that he is unreliable, but that he appears to be suffering from some sort

128

of eczema or psoriasis. He's another spirit that doesn't care much for the cold and is a lot slower to react in cool conditions.

The name Praseodymium comes from a Greek word meaning *green twin*, the green being his aura, the twin is not actually a twin spirit but rather his close associate and comrade, Lady Galåscor, the elemental of Neodymium. If you have water in the magic triangle, his aura will become yellowish green.

This elemental can teach you how to become patient, how to wait and not waste your energy trying to use force to obtain something for which the time is not right. He also has dominion over late bloomers, those who mature late in life and those who attain success later in life. If it seems that you have waited all your life and success has eluded you, Lord Genlascor can be a great help in this regard. Also, Lord Genlascor has dominion over character actors, those actors who do not take the lead role but are none-the-less so very important to giving a production life. He is said to have a high-powered magnetic personality and can teach one much about standing out, even when one isn't directly or is rarely in the spotlight, while still contributing to the success of the production. How one can shine without drawing attention to ones'elf and how one can exude great power and attraction while actually doing very little, the incredible power of keeping still, are some of the many lessons he offers. Thus, he is also a master of meditation and he gives the impression of great strength and endurance. It is said that Lionel Barrymore (look him up on the internet) was under his direction.

Lord Genlascor has a long association with Sweden and Austria and their peoples, cultures and magical traditions. He has a fondness for glass, and if you have clear glass in your triangle when he arrives his radiance is likely to give it a yellowish green coloration.

The gem peridot (also called olivine) can be used in the magic triangle to enchant him. It is said that peridot increases adventure in one's life, although it may also increase the challenges one faces. (If you are hoping Gandalf will show up at your door one day and invite you on an adventure, place peridot near your threshold.)

ॐॐॐ

Neodymium, Lady Galåscor
(pronounced gay – lah - score)
Lady of Magnetism
Chant: "I draw all I wish most easily, my desires now
all come to be."

Lady Galåscor is a powerfully attractive individual who can teach you everything you need to know about the powers of attraction, which include the power to repel those one wishes to keep away. While she has dominion over movie stars, both lead actors and supporting actors, she can also teach you how to attract anything you wish in life, whether you wish to be in the spotlight or not. She is known for having a great fondness for China and its peoples and is a master of the I Ching and Taoist Magic and can inform you of anything you desire in regards to these and other aspects of Chinese magical traditions.

She will generally appear with Lord Genlascor, who is her near constant companion, although, she is also known to hang out with the other lanthanides at times. She tends to wear dark silver gowns and while she is not necessarily pretty in a traditional sense, there is something about her that is incredibly magnetic. She exudes a reddish purple aura, and while you may not be able to see it, she emits a lot of infrared light. She is known to be able to shift her coloration according to the lighting you have in your magic room. She has a fondness of garnets and may be wearing jewelry that features that gem. But she also has a tremendous love of glass and crystals of all kinds, and any one or group of crystals can be used in her

130

summoning. Find a good book on the symbolic and radiant properties of crystals if you wish to use crystals to help indicate to her the type of energy and power you desire to gain or improve in you life. She is further known to support the use and development of hybrid cars, wind power and others forms of sustainable energy.

This elemental is also noted for being a matron of strippers and other dancers and performance artists, particularly those who turn up the heat, so to speak, as their performance draws to its climax. Additionally, she has a fondness of musicians, particularly those who use modern electric instruments. The name Neodymium comes from the Greek words for new twin, since this spirit is thought to be younger than Lord Genlascor, which may account for her interest in youth culture.

Besides China, she has also been known to frequent and favor the peoples and cultures of United States, Brazil, India, Sri Lanka, and Australia. Any of these traditions can be used to attract her. Since she has a propensity for triangles, she will feel quite comfortable being called within the magic triangle of manifestation.

She also knows a good deal about the history of the earth and particularly of the evolution of rocks, gems and minerals and can inform you about these. She is a bit of an amateur geologist. Also, she has sensitivity to volcanic activity and can tell one about a coming eruption and its likely magnitude thus she can also inform you of impending upheavals in the world or in your life. But once again, her greatest power is the power of magnetic attraction. She knows the secrets concerning attraction and can assist you to become permanently attractive. She is better at this than any other spirit.

However, do call her in the magic triangle. She has been known to have an artistic temperament, which is to say to flare up and have a tantrum at times. So keep that in mind. If she stamps her foot at you, don't panic, be patient.

ॐॐॐ

Samarium, Lord Sylmel (pronounced sill - meal)
Lord of Deception
Chant: "You do not see me, yet you do, I am other than
I seem to you."

Lord Sylmel is the patron of spies and all those who have need of using deception in their work. He is a master of protective coloration and camouflage and thus by extension has some influence over shape shifting. He can teach you how to fit in and even be a success without giving up, nor actually revealing, your true nature or principles unless you desire to do so. There is a tradition of magic where one does this very thing and weaves magic without ever being recognized as a magician nor appearing any different than the common man around one, this elemental can inform you of anything you wish to learn in this regard.

This spirit has a long history of serving as a catalyst, which is to say he is the welcomed stranger/visitor whose presence changes everything (see the movie or read the book *Cold Comfort Farm*). Elves have long used his services in hiding from the world of man and other, even more hostile, folk.

Appearing wearing a silvery copperish colored robe, filled with swirls and vertical lines, Lord Sylmel has strong magnetic power second only to Lady Galåscor; however, he has greater endurance than she, which is to say his charm remains even in heated situations, where others are more likely to lose their composure. At the same time, he has his limits and can be volatile under the right conditions, although the gaseous spirits of Argon have been known to have a very calming influence on him. Like Lady Rongra he is known to carry a lighter around with him, often clicking it on and off as a nervous habit. Whether he actually smokes or not, we do not know, although it has been suggested that he carries it to light smokes for his companions and others he encounters as a way of breaking the ice and doesn't actually imbibe himself. He is also known to have a strong attraction to and relationship with the Cobalt spirits.

This elemental is often referred to as a demon slayer. He is known in his radioactive phases for killing cancer and other intrusive destructive energies. The mineral dolomite, used for killing aberrant cells, can be utilized in his evocation. He is also known to have some power to control the fire/radioactive spirits so that their powers are used for good and not unleashed to wreck destruction upon the world.

He has a particular fondness for the traditions and peoples of China but also has been known to have a love of the magical traditions of the United States, Brazil, India, Sri Lanka, Russia and Australia. He is also a bit of an amateur archeologist and geologist.

Like a number of the previous spirits, he supports sustainable energy, particularly the development of solar power. He puts a great deal of energy toward research in the creation of solar powered aircraft. He is very much into developing the technologies for the decomposition of plastics and other waste products and pollutants. He also has a fondness for modern music, especially a love of the electric guitar. In many ways, he is a thoroughly modern spirit and very much carries a vision of a better and greater future for humanity.

Europium, Lord Nosor (pronounced no - soar)
Lord of Illumination
Chant: "Brighter now my life to be, a world of wonder
I now see."

Lord Nosor represents one of the least abundant elements in the Universe, so he is a very rare being. He is the Lord of Illumination and can help enlighten you or assist you in attempting to enlighten others, which is not always very easy. He can also increase your night sight, your ability to see in the dark, and heighten your personal radiance so your aura glows as is so often portrayed in pictures of holy people.

Like many of his kind, Lord Nosor prefers to dress in silver raiment, although in his case it tends to be subdued rather than shiny. He is noted for an Asian appearance and often wears clothing of the sort you find in Chinese historical dramas. He sometimes gives off a reddish pink aura and at other times a blue one. He is also known for being able to produce the *pure white light* that is so often associated with enlightenment. Yet unlike many of the other elementals just previous in this book, he is not very fond of heat. He much prefers the cooler climes, and functions extremely well in cold conditions. In many ways, he functions even better when it is cool. He is, however, very reactive to air and water, and even mineral oil, which is so conducive to helping many of the other spirits stabilize their manifest state, is only partially effective in his case.

He will never appear alone and often hangs out with the other lanthanide elementals and the raw earth elementals, although he is one of the only ones that are actually rare. Fluorine's Lady Leporn and Calcium's Lady Cahyrcor are totally infatuated with him and have been known to get very excited and blush when he is around and at times titter like schoolgirls. They think he is very special. Secretly, he has confided that he thinks that they are really more excited about being excited than really interested in him.

He is a student of both geochemistry and petrology and has a particular interest in studying igneous rocks, as well as volcanic activity in general. He has a fascination with magma and lava and can tell you more than you probably wish to know concerning these things. On the other hand, if you are interested in volcanoes, he's your guy.

China and its peoples and culture fascinate him, like so many other elementals. He really feels at home there. Surprisingly, he loves television and if you have a bright full screen television going in your magic room when you call him, that will be okay with him, although it may be hard to keep his attention as it tends to wander toward any active television set in his vicinity. If you could be playing some video that is symbolic of what you desire from him, this can be very effective in communicating your desires to him. Many folks of a spiritual nature dislike television, but Lord Nosor would very happily inform them how educational and enlightening television can be. He also feels comfortable under fluorescent lamps so those can be used as well.

The ability to detect counterfeits of any kind also falls under his purview. He's particularly adept at detecting counterfeit money.

ॐॐॐ

Gadolinium, Lady Anomal
(pronounced a - no - male)
Lady of the Shield
Chant: "Stand beside me straight and true, protect me
from what wicked do."

Sometimes called Our Lady of the Shield and other times a Shield Maiden, this warrior spirit can teach you how to carry your protective aura, your wards, wherever you go. Also, she can tell you all you wish to know about magical shields including those used as escutcheons bearing family crests and coats of arms. Eowyn in the *Lord of the Rings* would fall under her care.

It is one thing to have a protective barrier around your home and magic circle; it is another to have the power to sustain a magical orb of protective light about you anywhere you choose to adventure. This spirit can teach you how to establish and maintain such an aura so you are not only protected from the bombardment of negative energies of the

135

soul but also any attempts upon your psyché or physical body. However, this is not an introverted being. Part of the protection she has and offers comes from her extroverted nature. She associates and easily interacts with nearly all other elementals, and can tell you how making allies is far superior to acquiring enemies. At the same time, she is noted for targeting demonic activity that manifests itself as tumors.

Coming wearing silvery white armor, this elemental is very flexible and adaptable, and many say that she has a salty or earthy personality. She often has a greenish aura. She will never come alone but always accompanied by her comrades and fighting cohort. She is very encouraging and is known for heartening the spirits of the Iron and Chromium elementals making them less likely to break or give way under pressure. She especially favors emeralds, which can be seen on the front of her breastplate, but also other hexagonal crystals can be used in her summoning, such as benitoite, apatite and vanadanite. Garnet is another favorite gem. She does tend to get heated up by magnetic personalities, so if you wish to increase her attraction for you, have some magnets (three would be good) along the lines of or at the corners of your magic triangle.

She doesn't, however, care much for moist air. It tarnishes her armor, so a dry atmosphere for your magic room is much preferred for her conjuration. On the other hand, she doesn't mind water in your triangle, although if it is cold she will be slower to react than if it is heated, so keep that in mind.

She has an ancient friendship with the Finnish, Swiss and French peoples and their cultures and magical traditions. In modern times, she has founded homes in China, USA, Brazil, Sri Lanka, India and Australia. She is known for promoting nanotechnology. And like Europium's Lord Nosor, she has a fondness for television and, similar to other folks who are not exactly homebodies, she is perfectly fine with food offerings that have been cooked in a microwave oven.

Since she can obviously be a dangerous spirit, it is wise to call her within the triangle of manifestation.

> ### Terbium, Lord Zoncor (pronounced zone - core)
> Lord of the Voice
> Chant: "Resonant my voice will be, vibrant,
>   uttered powerfully."

Lord Zoncor can teach you the power of your voice. Not the use of words exactly, but the use of pure sound and vibration to affect the world. He doesn't actually give you anything you don't already have really, but actuates the power you have within you to give resonance to your tone so that it has a subtle and at times hypnotic effect on those you wish to influence (listen to the songs of Barry White). He can inform you how to use sound to navigate in the world, and how to use chants as a resonant power with which to communicate with the astral plane and the formative planes of being. He is the Master of the Voice and the Jedi Mind Trick. Saruman in the *Lord of the Rings* was said to have had this power.

He can further demonstrate how to use flat surfaces, such as glass windows or doors to turn vibration into subtle sound resonators. Since he has the ability to affect the "trichromatic" color perception of humans, this elemental has the power to use sound and magnetism to project a different visual aura to others thus in this way he teaches a form of shapeshifting, although one doesn't shift shape so much as shift other's perception of what one appears to be. This is a form of glamoury.

Wearing silvery white robes, a pink scarf and surrounded by a glowing green aura, Lord Zoncor will look very much like a blind man, often wearing sunglasses and carrying a long thin staff that he uses to tap about him. While he is a modest being, he is known to have a powerful influence on those around him and can teach you to have a similar effect with very little effort. He can help you learn to make your whole being hum with energy. He can further show you how to channel this energy through crystals in a stable manner so that it reaches out from crystal to crystal giving you a vast reach and power. He can also show you how to transform energy from, for instance, food to psychic electrical energy. He is also known for being very adaptable as a spirit.

The minerals of fluorite can be used in his conjuration, particularly those of the yellowish and orange variety. This spirit likes to associate with the elementals of Nitrogen, Carbon, Sulfur, Phosphorus, Boron, Selenium, Silicon and Arsenic, particularly when he is excited and feeling he is on a *hot streak*.

He has an ancient connection to Sweden and her peoples and culture, although these days he likes to spend a good deal of time in southern China. Like some of the elementals just previous, he has a love of television, although a magic room that is very colorful will also be attractive to him.

ॐॐॐ

Dysprosium, Lady Diculto
(pronounced dye – cull - toe)
Lady of Desirability
Chant: "You want me with a yearning bright, and hunger
for me day and night."

Dressed in dark silver, as so many of her kind, with a slight bronze tinge, this elemental will never come alone to your circle. She likes to hang with the crowd and always has others around her, although she may appear to be very reserved, inviting but shy. She frequently stands in the shadows. She sometimes appears to be a pure, virginal yet incredibly sexy and hot nun. She gives one the idea that she is unattainable and yet, with a sweet smile she lets you know that perhaps you are the exception.

138

In fact, there seems to be nothing that this spirit does that others cannot do as well. Except one thing. The name Dysprosium originates from a Greek word meaning "hard to get" and this spirit is an expert at what has been termed at times as *playing hard to get*. This spirit is the master/mastress of using reserve and mystery to create an aura of desirability.

If you have been selling yours'elf short, or others have been undervaluing you, this elemental can help you learn to increase your sense of s'elf worth or even more increase your worth in the eyes of others, not by putting yours'elf out and attempting to grab the spotlight, but by being reserved and holding back, although still appearing obliging. Willing but greatly hesitant. She can teach you how to do this without overplaying your hand, nor giving away the game. Others will come to appreciate you greatly and you, in turn, will also find a greater sense of s'elf worth.

This power stems, however, from incredible self-control. She does not follow every whim neither does she chase after every desire, but holds back and attains greater strength. You could say that she is very magnetic, but the truth is she uses the magnetism of others to draw them to her. This is a magical technique that she can explain to you if you are interested. On the surface, she appears very susceptible, but she does not easily yield to that susceptibility and this intrigues those who encounter her. She may appear, as we said, to be like everyone else and yet she stands out as exceptional. She appears soft and yielding and that is part of her great attraction.

Lady Diculto has a love of France and the French people and culture and a fondness for California, although she has also been known to reside in southern China and in Australia. She is very much into clean energy production, although she is in such high demand it is questionable that her talents will be able to be utilized in all the areas that desire her. She also has a very long and accurate memory. Anything you confide in her will be stored and remembered and she can help you improve your memory.

Garnets are a favorite of hers, and these can be used to attract her to your triangle of manifestation.

ॐॐॐ

Holmium, Lord Dutinu (pronounced due – tie - new)
Lord of Wizardry
Chant: "A wizard I would to be, the world to wander ever free."

This is the Lord of Wizardry, who can turn your magic wand into a staff, can increase your magnetic field, extend your attraction, and make it so you can, if you choose, like Gandalf, become an itinerant wizard wandering the world doing your magic wherever you deem best. Of course, one must be, as this elemental is, very adaptable to succeed in a life of wizardly peregrinations. He can further instruct you on how to absorb energy from others, particularly charismatic and magnetic attractive energy, without draining the other. It is rather like a transfer of fire. Your match will light ours and yet you still have your fire. This is, of course, different from vampiric transfer that drains the other of their energy. Here something is gained but nothing is lost.

Unlike Gandalf, however, this elemental wizard will never be found without his fellowship, companions and apprentices accompanying him. It is his nature to always work with others. Part of the reason he wanders is to visit friends but he never overstays his welcome. A sagely bit of wisdom that. He is sometimes known as Dutinu the Silver due to the silvery white wizard robes he wears. However, he has been known under some circumstances to wear a brown robe in order to disguise himself.

As we say, he is not inclined to be alone, but if you do summon him and he appears without companions, this will be okay as long as the air is dry. He's not fond of moisture or humidity and gets cranky under those conditions. And you don't want to make a wizard cranky, now do you?

While he will dress in silvery white, as we said, his aura will change according to the lighting. In daylight, it will appear bright yellowish, however, under artificial lighting conditions it will seem to be bright orange red. Your own eyes, under normal conditions, which is to say not in bright sunlight or the dark, will also tend to see this radiant orange red aura.

He is indeed a very charismatic being, and few come away from him without feeling that there was this *moment* between them when something truly magical occurred, or something subtle but powerful was transferred from him to them. It is as though he, without them being able to precisely say why or how, has altered the course and direction of their life, drawing them closer and more profoundly into the realms of magic and wizardry. When he is around Yttrium spirits, this magnetic quality is even greater.

He has an ancient relationship with Sweden and its peoples and mythic and magical traditions. He can tell you much of the magic of the Viking folk of that area. But he's also been known in modern times to settle for a while in China, United States, Brazil, India, Sri Lanka, and Australia and is familiar with all these magical traditions. In ancient China, he was known to associate with the wizard Bodhidharma who introduced Zen Buddhism to those folks. He has a love of the earth and the sea, but has little to do with the air, which is to say he is a practical and feeling wizard, but not much given to theoretical discourses. He believes in action not talk.

He does have a fondness for garnets, as so many of his close associates, but he also likes cubic zirconia and glass, particularly yellow and reddish orange glass. If you have yellow, orange, red or orange-red bottles and can create sun water, you can have that in the magic triangle to attract him, but keep the bottles closed. He's not into the water but rather the color vibration.

෨෨෨

> ### Erbium, Lady Ansomel
> ### (pronounced ane – so - mell)
> Lady of Starlight
> Chant: "My home among the stars I'll find, and
> to my destiny I'll bind."

This is the elemental of starlight and communication by light, thus she has dominion over the flash Morse code signals that naval vessels use, but also of the communications we receive by starlight from the Universe. By extension, she has province over smoke signals and signal flags and other means of communication that depend upon light and sight. Further, Lady Ansomel can teach you of the secrets of collecting and projecting light with and from your eyes to influence and communicate with others. These elves have found this to be a particularly effective method for dealing with those whom we can't communicate with directly, or who are simply closed to or incapable of reasoning. We soothe them and persuade them without them ever realizing how we do so or that we are doing so at all.

This is another extroverted spirit that loves to be around others of its kind, particularly the rare earth elementals. It is especially close to Yttrium, Ytterbium, and Terbium since they all grew up in the same neighborhood. These spirits like to be together and cannot be easily separated. Like many of her kind, she tends to prefer to manifest wearing silvery white clothes but has a fluorescent pinkish rose aura. She is known to have beautiful skin and wonderful teeth. She is also said to have a love of cheap jewelry and often appears wearing sunglasses. Those who have encountered her think that she does something to improve and increase their metabolic rate, although this, as yet, is just personal observation and reaction to her and it is unknown whether this is true or not. It may be that, like attraction or love, the effect is different depending on the individual. She is said to be a very magnetic being but at the same time incredibly adaptable, able to change herself as suits the environment she finds herself within. She can make herself seem the same as others, or opposite to them, depending on whether they are attracted to those who are like them or are challenged by those who are different and/or seem to reject them.

This spirit has an ancient relationship with Sweden and its peoples and mythology, although in modern times she seems to prefer to reside in southern China. She is known to have a love of colored glass, in particular art pieces made of glass, so if you have any art glass that will greatly attract her interest. She is also known to have an interest in clays, particularly those that are used in cosmetics and facials, or for healing purposes. She has a fondness for cubic zirconia so this can be used in her conjuration.

If you for some reason have any difficulty when summoning the elemental of Vanadium, you should know that this spirit tends to soften that elemental's stance and make her more willing to cooperate. Also, Lady Ansomel is known to be able to function even at extremely low temperatures. She is further rumored to have a very gentle and delicate touch.

While she is fairly safe to be around, she doesn't like dust, so be sure your magical room is well dusted prior to inviting her over.

෨෨෨

Thulium, Lord Rodcor (pronounced road - core)
Lord of Unexplored
Chant: "Beyond here lives a world unmapped;
  maybe a land where dragons happed."

---

*e*ver seen those maps that have at their borders, *beyond here there be dragons?* That is the spirit of this elemental. Lord Rodcor can take you to the realms

143

beyond the known into the dominion of the unexplored. He is a patron of pioneers of all sorts. If you wish to explore new realms that is to say realms that are new and previously blocked to you, this elemental can be of great help. Whether it is in the material world, the astral world, the spirit worlds, in this dimension or some other, Lord Rodcor can ease the way, show you the portal, and give you the travel guide that will help you succeed in your quest. He can teach you how to move from one parallel world to another, and how to arrive at the dimension you seek to explore. The notion of *you can't get there from here* is not known to him. This elemental will tell you that you can always get there from here, just not always directly.

Lord Rodcor is a very rare spirit and doesn't come to the Earth often, although he does have some residences here. He will manifest wearing silvery gray raiment, which tends to slowly darken the longer he stays. He has a green aura; however, under ultra-violet light he will appear as being bright fluorescent blue. He is not particularly dangerous, just a little hard to enchant. He has a strong sense of his own value. Yet, he is known for being soft and easy going if you do get him to appear. He is known to associate with the spirits of Erbium and Holmium and like many of his kin he has an ancient home in Sweden and is very fond of its peoples and mythology. He also has an ancient relationship with Iceland. These days he is known to frequent China, Australia, Brazil, Greenland, India, Tanzania, and the United States.

And as so many of his lanthanide kin, he doesn't like to be alone and it is extremely unlikely that he will appear by himself. He is, however, relatively adaptable in his solid state but since Thulium becomes volatile in liquid form, there are limits. He can go with the flow but he doesn't like to be forced to do anything.

He does have a fondness for rivers and likes to hang out on their banks or in their sands when they are shallow. So if you can summon him near a river, or in a shallow or dry streambed, it would increase your chances of attracting him. He is known to have an interest in the medical field, particularly surgery, and promotes advanced surgical methods. He is also known to have some weather-working ability, as well as an interest in military matters and can teach you a bit about defending yours'elf in strange climes. He has powerful eyes and x-ray vision. He's another spirit that doesn't care for dust, so be sure to have your magic room purified before summoning him. If you do your conjuration by a river, fresh air and a clear day are best.

ॐॐॐ

Ytterbium, Lady Vancor (pronounced vein - core)
Lady of Precision
Chant: "Exact, precise and without error, my magic true
    and nothing fairer."

In the late 1960's movie *Simon King of the Witches*, the magician Simon asks an eager young protégé how one can expect the gods to answer one if the magician doesn't even pronounce their names correctly. In Jonathan Stroud's great *Bartimaeus* book series, a magician's undone because he summons a demon but doesn't have the wards around his magic circle written correctly. The idea that one's magic, like some chemistry formulas, must be utterly exact or one risks failure, perhaps death, is not an unusual idea, although elven magic and the spirits we interact with, for the most part, are far more flexible. In enchantment, intent and style is more significant than detail. However, there are places in life and magic where precision is very important and this elemental can be of great assistance in those situations for she is the Mastress of Precision. If you really need to be there on time, or to have your timing exact, she is the spirit to call. (However, don't mistake her for Yttrium. Get her name right; she will greatly appreciate it.) Some folks say that *time is money*, but she will tell you it is also power and effectivity.

This elemental has childhood associations with Yttrium, Terbium and Erbium, who all share a love of Sweden and its peoples, magic and mythologies, although these days she's commonly seen in China, the United States, Brazil, Sri Lanka and India. Like most of the rest of her gang, the Lanthanides, she wears the gang colors of silver or silvery white. You may notice a bright emerald green aura around her. She always wears a watch and you may find her checking the time if you seem to be babbling or imprecise with your evocation. Make her conjuration short and succinct. Know what you are going to say ahead of time; and don't waste hers. She can be quite testy if you aren't precise and clear in your evocation. Also, be sure to summon her within the magic triangle, as she's been known to irritate the magician's eyes and skin if one does not do so.

Some have quipped that she is not exactly easy on the eyes and has that stern, anal-retentive librarian or nurse look about her, although she is clearly a brilliant being. She is said to have great teeth but since she rarely smiles, it is hard to tell. However, that is just her appearance and the first impression one may get. If you do enlist her aid, know that she can be quite soft and adaptable, especially if you follow her suggestions. Otherwise, she just thinks you are wasting her time and, as we've said, she has a thing about time.

If you have some cold water in the triangle, this will tend to slow her down a bit. She is also known to be partial to face centered cubic crystals and thus pyrite, sphalerite and galena can be used to attract her. Pyrite has the property of helping you to harmonize with her. Sphalerite promotes empathy between the two of you and Galena helps you to recognize the truth when you hear it so you cannot be fooled. Also, she is hard to separate from the others in her crew; so don't expect her to show up alone. And, like some of the others of her crowd, she doesn't care for dust. In fact, she hates it and can explode into a tantrum if your magic room isn't clean. So purify before you conjure.

ॐ ॐ ॐ

Lutetium, Lord Ariscor (pronounced air – rice - core)
Lord of the Visitation
Chant: "Come to me from far away, enlighten me
        this very day."

Lord Ariscor prefers a slightly darker silver tone to his clothes than most of the others in the lanthanide gang. He has a white, almost clear, aura and has an aversion to moist air. As you gaze at him, it may seem like the air around him is undulating, like a heat mirage. By all means, call him under dry conditions.

He is usually perceived as a short, compact, tough guy. He looks hard and he is, although he often appears to be nervous or edgy, or some might say, slightly unstable, however, he is not known to be particularly dangerous unless he loses his temper. Cool water in the magic triangle can serve to slow his reactions. And like some of his associates, he doesn't like dust; so keep your summoning area clean and dust free.

He is the last of the lanthanide elementals but also hangs with the transition metals, which are another gang. He is a bit of a go-between. He can be found associating with nearly all the rare earth elementals. He is known to have a very good memory.

He first revealed himself to Austrian, American and French alchemists, who noticed him in a group that tended to hang out with Ytterbium's Lady Vancor and realized that he was not merely one of her followers or assistants but an elemental himself. He has an association with the constellation and Greek mythological figure of Cassiopeia. This association is a reminder to be humble however great, powerful or

147

handsome/beautiful one might be. He is frequently found with the elemental of Yttrium. He also relates to the star Aldebaran whose name, coming from Arabic, means *the follower*, because even though it is one of the brightest stars that we can see from Earth, it rises after the Seven Sisters or the Pleiades. Again, it suggests that no matter how great one is, modesty is a vital power.

This elemental is the liaison to all spirits that visit us periodically via the meteors, comets and meteorites. He is also by extension the lord of omens and oracles that come from these visitations, these messengers from the stars (note that Mark Twain was born and died in the year of Halley's comet which comes every 75-76 years). If you wish to connect to these spirits, and particularly if you have seen or dreamt of meteors or comets lately, then this is the elemental who can be of assistance. Also, if you did have a dream about comets, this spirit can aid you to interpret and understand it. He also has some healing skills but mainly dealing with tumors of the endocrine system.

He has an especial fondness for Paris, particularly the oldest parts of the city. He has an ancient relationship with the Gauls of that region. In modern times, he can be found, as is the case with most of his associates, in China, United States, Brazil, India, Sri Lanka and Australia. He has a fondness, like so many other elementals, for garnets.

Hafnium, Lord Dencor (pronounced dean - core)
Lord of the Dragons
Chant: "Of Dragons' power I would know,
      by fiery breath let wisdom grow."

Lord Dencor is the Lord of Dragons, or more commonly called a Dragon Lord. He can tell you of anything you wish to know concerning dragons, their lore, powers, legacy and magic. By extension he can inform you of snakes, lizards, dinosaurs, worms and everything else that is connected to dragons (see Marie Brennan's great Memoirs of Lady Trent series starting with *A Natural History of Dragons*). He is another elemental whose being was prophesized and like dragons, he can breathe fire, pouring out jets of flame up to a hundred feet, so by all means, call him within the magic triangle. He is said to have an incandescent aura.

He will appear wearing shiny silvery gray armor made of scales, which makes his armor very flexible, and makes him resistant to attack. Though silver colored, it may seem to have a slight greenish tinge to it. He is known to have an ancient relationship and love of the city of Copenhagen in Denmark and a fondness for the ancient Danes. These days, he tends to reside in Brazil and Malawi, and Western Australia.

He has a very strong relationship with the elementals of Niobium, Titanium, and Tungsten. He also has a longstanding relationship with Zirconium. These two are so close that they are often thought of as soul mates, although Lord Dencor gives one the impression of being strong and solid whereas Zirconium's Lady Fadymcor seems a bit more delicate. None-the-less, being the Mastress of Wards, she is truly as powerful as he is. And you can be certain that if he manifests, she will be right there beside him. They are nearly inseparable and he tends to follow her everywhere.

This elemental is known to have a fondness for zircon crystals so these can be used to attract him. He finds the mineral eudialyte very attractive and also rutile quartz, although this latter because Lady Fadymcor likes it. Because of an association he has with squares, draw a square within the triangle; this will make him more at ease and more stable. Enchantment is the key to attaining his aid as well as Lady Fadymcor's. Traditional magicians who think that they can summon them and order them around are likely to encounter their explosive natures.

Because he understands dragons and has the power to control them, he can inform you of the dragons that exist in the world in human form, that is to say as spirits in disguise, and how one may best deal with these incredibly powerful beings, who have, as their legends suggest, a tendency to hoard wealth.

అఇఇఇ

Tantalum, Lord Harala (pronounced hair – ray - lah)
Lord of the Cosmic Carrot
Chant: "From desire I'll be free, and from this
all things come to me."

The Cosmic Carrot is the reward that is dangled in front of you by the Universe in order to entice you toward enlightenment. Some magicians, both of the light and dark sides, have learned to use people's desires to manipulate them toward doing what they wish, often suggesting that they can fulfill one's desires or otherwise give one something that frequently they cannot or have no intention of doing (listen to most politicians). Whether one does this for the individual's benefit or to their determent determines the ethical nature of this magic.

Lord Harala can teach you everything you wish to know about using rewards as an impetus for achieving your will and affecting the behavior of others. He can also teach you how to resist such temptations so you are not manipulated due to your

150

desires but become truly free. He can also point out to you the reality that positive reinforcement is more powerful when it is intermittent (as the Behaviorists have demonstrated) than when the reward is given every time the individual performs the chosen task. This, in a sense, furthers the notion that *goodness is its own reward* or the action is rewarding in and of itself.

Named after Tantalus, the Greek mythological figure that was punished by being eternally tantalized and unable to fulfill his desires, this spirit will manifest wearing shiny dark blue-gray clothes. Since Tantalum is highly resistant to corrosion, we come to understand that this elemental has overcome the desires that were previously used to torture and manipulate him, and as we said, he can show you how to master your own desires. This is one tough spirit and no matter how much one may try to tempt him or pressure him, only enchantment will bring him around.

This spirit can teach you how to make yours'elf stronger and in many ways nearly unbreakable. He can also show you how to store up energy temporarily in talismans to be released later (read the *Dresden Files* series where the wizard Harry Dresden accumulates energy in a bracelet that he uses in emergencies). He can also show you how you can draw energy from mobile phones, DVD players, video game systems and computers or use them to project your magic through the electrical system (see Taylor Ellwood's *Space/Time Magic* for more on this.). He very much promotes modern technology, especially bionics and he strongly supports recycling.

He will always appear with his companion and kin Niobium's Lady Unos. And he favors the mineral tantalite that helps one change depression into passion, although be careful if you tend to suffer from manic depression for you could end up suffering from pure mania. Since Tantalum is more rare than Gold in the Universe, you can understand that this elemental has a sense of being both unusual and exceptional. He is a very adaptable being and can be soft and easygoing or hard and brittle as the circumstances demand. He is a pretty

151

independent spirit and you can't force him to work with other elementals against his will.

Like many elementals, he has an ancient relationship and interest in Sweden but also in Germany and Switzerland. He currently has residences in Australia, Thailand and Malaysia, although it is said he is acquiring property in Saudi Arabia, Egypt, Greenland, China, Mozambique, Canada, the United States, Finland, and Brazil and even more land in Australia. Naturally, you can use the mythologies and magics of these cultures to help attract his interest or use him to influence and affect these lands and peoples.

Also, this elemental is in high demand and there are dark sorcerers who are attempting to dominate him and keep him for themselves who may block your attempt to summon him. Have your wards strong, not against him but those who would interfere with your magic. For his own part, he is very compatible with most human beings.

Tungsten, Lady Bramyl (pronounced bray - mill)
Lady of Sorcery
Chant: "I will to penetrate the heart of every matter
from the start."

Tungsten comes from a Swedish word that means *heavy stone*, but since they use this name for the mineral scheelite, the Nordic folk call this element Wolfram. Thus this elemental has an association with scheelite, a mineral that can be used to balance the first charka, which is to say help one obtain stability and safety in the material world. Wolfram (volfram in Swedish, also the name of the evil corporation in the *Angel* television series), which seems like a combination of wolf and ram, actually means wolf's froth or saliva and thus also connects this spirit to the wargs, the

152

enormous wolves (vargr) of Norse mythology, fantasy books and movies.

But Lady Bramyl's main power and knowledge concerns the magical discipline of Sorcery. She can inform you of anything concerning the art of seeking the source, the underlying causes, energy and powers that create and move all things. She can help you illuminate the dark areas of your being or your life, see into the occult realms of manifestation, help you make strong and enduring bonds with the spirits of those realms including finding true and faithful allies, and help shield you from the dangerous powers and emanations of those planes of being.

This spirit will appear wearing steel grey robes and will seem to be, and is, incredibly strong and compact. It is said she has a steely, penetrating gaze. She may radiate a yellow aura. She frequently wears metal jewelry, although she has also been known to have a love of ceramics. She may appear wearing a knife. Though she is a sorcerer there is a warrior spirit about her (see Castaneda's Don Juan books for more on this connection between warriors and sorcerers). She can be brittle at first and hard to get to work for you, however, the secret of this elemental is to keep calling her. She likes to be courted and the more you call her and enchant her the more malleable she becomes. So don't give up just because she doesn't reveal everything to you all at once. And she is not merely an intellectual. This Lady walks her talk, knows from experience, can take the heat and pressure and has endurance greater than nearly any other. She has been to the darkest areas of the Universe, has walked through its hells and has come out again.

This elemental has an ancient relationship with the Basque peoples of Spain, France and Portugal. She also has a love of the Dartmoor region of Britain. Besides scheelite, the mineral ferberite can be used in the magic triangle. This mineral helps one adjust to new situations so this is beneficial both to the manifestation of this spirit and for the sorcerer as sHe explores new realms. She is known to have a thing for fast cars (she loves speed), a love of fishing, and playing darts. Also, she has a fondness for stringed musical instruments. Keep all this in mind when conjuring her.

వావావా

Rhenium, Lady Radencor
(pronounced ray – dean - core)
Lady of the Skies
Chant: "Quickly now and evermore, I will travel
and explore."

While many elementals are masters of various aspects of the development of personal power, Lady Radencor is the patron and Mastress/Master of jets and travel by jet aircraft. Thus she also has dominion over pilots, co-pilots and all other things having to do with traveling by jets. Timothy Leary said that, "Mobility is Nobility", and this surely applies to those who jet around the world on a regular basis. Therefore, the *Jet Set* also comes under the purview of this elemental. If you desire to join the *Jet Set*, to increase your ability to travel by jet, or to know anything that has to do with jet travel, this is the spirit to conjure. Remember that time moves slower, or really you age less quickly and clocks move more slowly, the closer you approach the speed of light. If you are interested in becoming immortal, this is a good place to start.

This elemental will manifest, as so many others of metallic disposition, in shiny silvery white. It is said that her eyes beam light like a camera flash and can temporarily blind you. You may wish to wear sunglasses when summoning her. She has a resemblance to Manganese and Technetium elementals; however, she tends to associate more with the Copper and Molybdenum spirits. Lady Radencor is a very rare being and, of course, being part of the *Jet Set*, rather wealthy. She is another

154

being whose existence was prophesized before she actually chose to reveal herself.

She has an ancient association with the Rhine river that begins in Switzerland, passes by France and Germany and ends in Denmark. She thus has association with the Rhinemaidens (water nymphs), whom legend says stole the Ring of Power from the Alberic the Nibelung, Dwarf Lord (or dark/unseelie elf) of the Underworld (see Richard Wagner's opera cycle *Der Ring des Nibelungen*). Note that Oberon is the French translation of Alberich and is used for the name of the "King of Fairies" in a *Midsummer Night's Dream* (alb meaning elf in old German). Therefore, this elemental has dominion over water nymphs, sea sprites and mermaids, and from that by extension province over sea travel, particularly cruise ships, as well as air travel.

In modern times, this spirit can be found mostly in Chile, although she has recently been spotted on Russia's Kurile Islands. She also has association with the peoples and cultures of United States, Peru, and Poland. She is known to have a strong and symbiotic bond to the elemental of Tungsten.

Osmium, Lady Luwyn (pronounced lou - win)
Lady of Soul Mates
Chant: "Destined we together be."

Lady Luwyn rules the coming together of Soul Mates. If you have not as yet found your soul mate or mates, she can surely be of great help; however, she will also inform you that in order to do so you must necessarily come together and work out your relationships with everyone with whom you have a karmic debt. All these debts you owe must be cleared away before you find the one or ones that are truly meant for you. And since Osmium comes from a Greek word meaning smell, you can be certain that you will find your soul mate not by any logical effort but by following the instinctual, perhaps even primitive or more accurately the primal, directives of your soul.

Lady Luwyn can also help you find those who will assist you in connecting with your soul mates, those individuals with whom you have a spiritual destiny to work together toward the great fulfillment of the magic, the great work, the mastery of yours'elf, your magic and your life energy in harmony with your kindred and for the betterment of all. If you are not really ready for your soul mate(s), do not despair; she can help you find others who will fill in, so to speak, until the right time comes. And since soul mates are by definition those who are right for your soul, she can also inform you of those who are wrong for you, those individuals that tend to infatuate you but are not good for your spirit or soul.

Manifesting wearing bluish white clothing with a sort of gray aura, you may notice that this spirit has wings of the bat-like demon/gargoyle/pterodactyl kind (as opposed to the butterfly/dragonfly/faerie type, or the bird/angel sort). You may see an old fashion fountain pen sticking out of her coat pocket. She also has a reflective, mirror-like ability, so if she shows up looking like a mirror image of you with wings, don't be shocked or put off your game.

She is part of the Platinum group; that elite clique of rare and wealthy spirits who tend to hold themselves in high regard and expect you to do so as well. They all reinforce this opinion of themselves and each other. And don't let their shiny exterior fool you, they can be very hard spirits. At the same time, it should be said that this really is one tough spirit who keeps smiling even when the heat is on. She is not easily fazed, although she can be a bit brittle and touchy at times. Be polite, courtesy is always appreciated. She may arrive with spirits of Platinum and Nickel accompanying her. They commonly hang out together. And this is really best, as she has a very toxic disposition when she's alone.

As a sideline, she has an interest in and great knowledge of Cretaceous–Paleogene boundary that marks the extinction of the dinosaurs and can tell you of that period and the disappearance of the dinosaurs. She has a fascination for pre-Columbian art and also a keen love of old fashion wax and vinyl musical records.

She finds the city of London very appealing and spent a good deal of time there in the past but these days she is found mostly in South Africa, near Norilsk in Russia, and the Sudbury Basin in Canada and has a fondness for the peoples of these areas and their traditions.

While, she can be dangerous, you may put a bowl of polyunsaturated vegetable oil, such as corn oil, in the magic triangle and this will serve to calm her and reduce the friction between you.

ॐॐॐ

Iridium, Lady Britonme
(pronounced bry – tone - me)
Lady of the Rainbow Bridge
Chant: "Though we be of different tribes, I sense we
share quite common vibes."

Named after the Greek Goddess Iris, the goddess of rainbows, this elemental is a spirit of the Rainbow Bridge that in Norse mythology is called Bifrost and connects the world of man with the world of the gods. If you wish to connect with the Divine realms of the gods (the Shining Ones for the elves), not just one god but the heavenly realm, this is the spirit to summon. This elemental can also tell you of nearly anything regarding rainbows in actuality and legend, and because of the association between rainbows and leprechauns those fae folk fall within her area of expertise as well. Any relationship you have or desire to have with the leprechauns, including the accumulation of gold, can be mediated by this spirit.

It has been suggested that Bifrost means "shimmering path" and this connects the rainbow bridge with the Shining Ones, and also with the path to divinity within ones'elf. Since the rainbow is used in modern times as a symbol of unity among diverse peoples, a very Aquarian ideal, this spirit can further help you make friends and find allies among those who are very different from you, who are connected not by race, religion or nationality but by vibration. This extends to the idea that this spirit can help you link to other species, aliens, and inhabitants of various realms that are incredibly different to our own. And because of this Aquarian aspect, any correspondences relating to Aquarius can be used in this elemental's summoning. If you do conjure her, know that she has a catalytic effect on those with whom she interacts; so you can be sure your life will change, particularly in regard to an increase in pressure to move you toward the realization of your divine s'elf and the divinity in others. She always brings a message of hope and renewal from the divine realms.

This spirit will manifest wearing silvery white garb with a slight yellowish tinge and a shimmering, rainbow creating, aura. She may reveal her bird/angel like wings to you. It may look as though she is standing in the midst of mist on a sunny day. She has a natural affinity to Platinum and is one of the Platinum crowd, and like most of the others of that clique, she is a very rare being. She spends more time in space, particularly riding meteors, than she does on Earth.

You should know that this is the spirit, along with Osmium, that was sent to the Earth 66 million years ago to end the reign of the dinosaurs. This opened the way for the development of ape/man and other species and channeled the dinosaurs into their current bird manifestations, which actually was a spiritual promotion for them. Also, this spirit took part in the formation of this planet, having a great influence on its magical and spiritual development when it was young and still molten. She is tough, can take the heat and is nearly immune to corruption.

She has a long-standing relationship with the Ethiopians and the South American cultures, particularly the early native tribes of Columbia. She is not a particularly dangerous spirit, unless she in fact comes bearing a message from the divine realms that may radically change your life. But know that even if your life is disturbed by her arrival the long-term outcome will always turn out to be a blessing for you. Her motto is: Everything happens for the best.

ન≈ન≈ન≈

Platinum, Lady Latyna (pronounced lay – tin - nah)
Lady of the High Elven
Chant: "I would my spirit higher be, an elf
of great nobility."

This elemental has association with the High Elven folk, the noble folk of elven kind who are by our very natures, of royal and noble being. All elves are considered nobility and yet among us there are those who are even greater, more spiritually developed and an example to all of us. If you would communicate with, associate with, or aspire to become one of these, this is the spirit to enchant.

Dressed in silvery grey white, this spirit is adaptable yet deep, stable yet flexible, and exudes a sense of profound presence. One feels that something truly amazing and wonderful is occurring when Lady Latyna appears. She will probably come wearing Platinum jewelry. She is said to have a very bright and charming smile. She is known to have a fondness for rivers and if you can evoke her near a river that would promote her appearance. She is not a danger to you, although she, like the rest of her crowd, holds herself in high regard, none-the-less, you don't need to summon her within a triangle. However, if you commonly use salt as a protective ward around your magic circle, don't. She would find that insulting and under those conditions she can be a bit testy. On the other hand, she hates demons and will help you banish any that may hang around your circle, so don't worry, she can be trusted and she's on your side. She can also help you convert your negative aspects into positive values, your weaknesses into strengths. It is said she will put an indelible mark upon your soul and spirit; a sigil that will help signify her approval and gain you access to and power upon the higher planes of manifestation, especially the elven faerie realms.

She is a very modern spirit in the sense that she supports the use of electrical gadgets. She is also known to have a love of the art of photography and etchings. She is also said to like porcelain.

It is possible that Platinum or an alloy thereof was what Tolkien meant by Elven Silver, and in fact, the name of the element comes from the Spanish and means "little silver". And as we pointed out, Lady Latyna heads her own clique, the Platinum crowd. She also has association with the Nickel and Copper spirits.

She is said to have a relationship with the pre-Columbian Ecuadorian natives, although these days she can often be found in South Africa and in the Ural Mountains of Russia. Furthermore, she has a relationship with the Ancient Egyptians, although it is rumored that she appeared to them in disguise. Spain, Panama and Mexico were also favorites of hers. She has further been known to spend time in Alaska and Canada and from time to time has been spotted in Montana in the U.S. and in India. Any of the mythologies or magical traditions of these cultures can be used to appeal to her as she has a keen interest in all of them or you can use her power to influence these cultures, peoples or geographic regions.

Since Platinum is more abundant on the Earth's Moon and on some meteorites than on Earth, Moon correspondences can be used in her summoning as well. Thus if you call her under the Moon by a river, you increase your chances of success. Also, she is noted for being slightly vain. If you have a mirror in your circle that she can use to check hers'elf out when she comes, it would be most appreciated.

❧❧❧

Gold, Lord Urm (pronounced your-m)
Lord of the Common Weal
Chant: "A people great and strong we'll be,
each finds success and ecstasy."

This is the spirit of material wealth, however, it is also the elemental of the Common Weal, the welfare and wellbeing of the people, also known as the Commonwealth, wealth used to uplift a nation, tribe, and society. This spirit will aide you to become wealthy, if that is your desire, with the caveat that you must use a good portion of that wealth to help further the wellbeing of the realm and its people. Would you be noble? Would you lift yours'elf up both materially and spiritually? Then you must assume responsibility for all of your kin. This elemental can help you do that. Can reveal to you the secrets of wealth and the wisdom of its proper use and distribution.

Unlike most of its kindred, this spirit will appear wearing reddish yellow clothing, instead of the variations of silver that are common among most of them. You may see a greenish aura. His clothes are often embroidered with golden thread. He is a unique individual, but part of what makes him unique is his drive to help and uplift all his others. He is quite easygoing and very malleable as a spirit. And while he has a family relationship with Silver and Copper, he is not afraid to come alone in response to your summoning.

He is known to love mountains, particularly caves, and also rivers. Some say that you can see his prominent veins when

161

he manifests. He is a very ancient spirit and was around when this Solar System was formed and watched and participated in the birth of this planet. His parents were neutron stars (great Shining Ones). He is often referred to as the *King of Metals* so you may find he is wearing a crown, and like the king that he is he lives for his people, for his land (the king and the land are one). However, unlike worldly kings, he is not easily corrupted. Know that in calling him you are essentially saying: let me be one of your people. Shed your wealth on me and I, in turn, will use it to help others. He is also a very calm and calming being and counteracts any tendency for those in his vicinity to get inflamed. Also, he is not a dangerous spirit, in fact he is very protective of those under his care, unless you're greedy and that really is your problem, not his.

His name in various languages means *to shine*, *to glow*, but also means *the dawn*, since his spirit brings the dawn of a new age. The greatest era on Earth is often referred to as the Golden Age. The highest values called the Golden Mean. He has province, by extension, over expensive jewelry and over money, particularly coinage. Interestingly, he has a keen interest in computers, particularly in their ability to unite the world.

Ancient shamans were known to use gold as a medicinal treatment, so this elemental also has a long-term relationship with shamanism. It is possible he could appear in his shamanic rather than kingly grab, however, that would probably depend upon your intention in conjuring him. He is the shaman king.

In the ancient world, he has association with nearly all peoples and cultures. In modern times, he can be found frequenting China, Australia, the United States, Russia, and Peru. Also South Africa, Ghana, Burkina Faso, Mali, Indonesia and Uzbekistan. This elemental gets around and is still highly valued although few who desire him understand the principles that are needed to successfully utilize his great intelligence.

ॐॐॐ

> **Thallium, Lord Poscor (pronounced posce - core)**
> Lord of Druidry
> Chant: "The wisdom of the trees I'd learn, of plants
> and herbs and magic fern."

*U*nlike the two previous elementals, who are amiable and friendly for the most part, this is a dangerous spirit, so by all means summon him within the magic triangle while you are in your circle of protection. He particularly doesn't like rodents or insects; so as we've said for Barium's Lady Sylwima, don't have your pet hamster in your magic room. But he's also rather indiscriminate when it comes to it, so your pet bird, rabbit, dog or cat could also be in danger.

Thallium comes from the Greek meaning green shoot or twig, and thus this is the elemental of Druids, those wise folks who could be healers, harpers and magicians of various arts (see Grave's *The White Goddess*), but could also be deadly if need be. They can use herbs to heal or to poison. By extension, this elemental is related to the Sylvan Elves particularly in their aspect of defending the forests and the wild places of Nature.

This elemental will never appear by his self but will always have fellow druids accompanying him. He generally wears gray robes with a bluish tinge but has a bright woodland green aura, like moss after the rain. He will be bald on top, although he may have long hair on the sides and back of his head. He is sometimes mistaken for the elemental of Lead, but he is most often found associating with the elemental of Potassium, although he can also be discovered occasionally with the Copper and Zinc spirits. He has the ability to see in the unseen realms, or that is to say to see spectrums of being that common folk cannot perceive and can help you learn to do the same. He is known to have a fondness for glass, although that may come from his association with beakers, test tubes, and jars filled with potions.

163

Historically, Thallium has been called *the poisoner's poison* and the *inheritance powder*. It is elven lore that among abused women, held in bondage and oppressed by men, this spirit is known as the *housewife's benefactor*. Magic can be used to curse or cure, to do good or ill and it is up to the individual practitioner to use it wisely. This spirit can inform you of the appropriate use of his powers; although, it is obvious these powers have been abused in the past.

Curiously, despite what some might think considering his potential for being dangerous, he is a soft and yielding being and rather sensitive. He doesn't like to be slighted. Honor him. Not because he's dangerous but because he's earned it. Also, he is soothed by oil, so you can have that in your magic triangle to make him feel comfortable. Iron pyrite can also be used in his conjuration. This helps one find harmony with beings of other levels of manifestation. He is known to have a love of the seas, so doing his evocation on a beach can heighten your chances of success, but naturally a forest conjuration would be equally productive. If you wish to have him stay for a time, the pigment Prussian Blue, sometimes called Berlin Blue, absorbs his energy and limits his toxicity. You may wish to paint your magic triangle with this color.

He has an ancient relationship with Macedonia and thus by association with Alexander the Greek and his father Phillip of Macedonia, however, his real link comes from the Celtic culture of that region that invaded after the fall of Alexander. He can be a powerful and magnetic being and can teach you how to draw energy and power from the ocean or the forest, or any growing thing (he has an especial love of vines), and use it for your magical purposes.

> ## Lead, Lord Emni (pronounced eem - nigh)
> Lord of the Military
> Chant: "Ready, strong and valiant be, protected,
> we are ever free."

This elemental is doubly magic and knows all that anyone would wish to learn about the military and, by extension, all hierarchal organizations, about strategy, tactics, logistics and anything else on that subject you may wish to know. He can also tell you about the magic of numbers, particularly the importance or at times lack there of, of superiority of numbers. He is a very ancient spirit and is also associated with and has province over plumbers, those who plumb to the depths of things, of dwarves as cave dwellers and miners, and particularly of sub-dwarves, which is to say half dwarves, hobgoblins and others who prefer the dark recesses to living in the light. Anything you wish to know about these folks, you can ask him. He knows both offensive and defensive magic and can particularly shield you when you are dealing with the radioactive/fire spirits.

He is, as you might expect, a dangerous spirit and should be called within the magic triangle of manifestation. He will appear to be soft and malleable, but remember that is in part because of his incredible confidence and as he will tell you one must be extremely adaptable to respond to ever changing battle situations. If you have to deal with the military, with top down corporations, with hobgoblins or plumbers, he can tell you the spells that will give you influence over them.

His raiment will appear bluish white when he first appears but will soon turn to a dull grayish color. He has a relationship with ancient China, Greece and Rome all of whom utilized his services. The Native Americans of the Missouri area also knew of this spirit and availed themselves of his aid. These days he is frequently found residing in Australia, China, USA, Peru,

Canada, Mexico, Sweden, Morocco, South Africa and North Korea.

Alchemists considered him to be the oldest of the metal elementals. And of course, their goal according to popular legend was to transform Lead into Gold, that is to say to transform military might, and the secrets of plumbing, into an advanced civilization. This surely was the goal of the Chinese, Greeks and Romans who are all noted for their advanced cultures in the ancient world. (There is just something about indoor plumbing and running water that gives one a feeling of superiority.) He commonly allies himself with Antimony, Copper, Cadmium, Tin, and Tellurium, thus increasing his strength and power. He knows that military might isn't just about being strong in oneself but strong in one's alliances. He is also commonly found with his friends the Zinc and Silver elementals.

He has a love of face-centered cubic crystals and minerals, such as pyrite and galena and these can be used in his conjuration. Also, he is partial to the minerals cerussite and anglesite. And, despite his knowledge of, and power concerning, hobgoblins and others of that type, he loves the sun. So don't think that he is a hobgoblin himself, he is not; he merely has the power to shed light on their workings. He's studied them for ages, and has magical influence over them. If you need to defend yours'elf against any of these, he can show you how to do so successfully. He can also help you to stabilize yours'elf, especially if you tend to be scatterbrained or a bit *airy-fairy*. He can help you get organized. He is also known to be a supporter of recycling and solar energy.

The planet Saturn is the astrological sign that is associated with Lord Emni, so the correspondences for that planet can be used in his conjuration. He further has a fondness for sailing. Also, he is known to love churches, stained glass, and organ music. Gregorian chants would work well in his evocation. Furthermore, he has a love of ceramics, so if you have any ceramic tools in your magic, that can be of help, particularly if they are handmade. Keep all these things in mind when you plan his conjuration. Let your intuition be your guide. Don't forget, while he is stable, he's also a serious and dangerous spirit.

ঽ৽ঽ৽ঽ৽

Bismuth, Lady Harwi (pronounced hair - why)
Lady of the Five Pointed Star
Chant: "Within my realm, secure and safe,
    my will fulfilled with total faith."

Lady Harwi wears silvery white clothes with a pink tinge and will be noted for her metallic iridescent rainbow aura. She will undoubtedly be wearing makeup and most probably pearlescent eye shadow. She has a resemblance to Arsenic and Antimony, particularly in her attitude. They all use intelligence rather than force to get what they desire. Physically, she is commonly mistaken for Lead and Tin. Lady Harwi is very independent, has a keen mind and may show up without companions. She will most likely question you about what you desire. You may feel like you're a lawyer before a Supreme Court judge. Also, she is known to be dyslexic so give her a moment to adjust. Some think she is dense and unintelligent, but that is a mistake. She merely needs to concentrate to interact with the human world.

This elemental rules the five-pointed star, which is to say the four elemental states and their interaction with the Divine Magic. The five-pointed star, among others things, represents Mankind (while the seven pointed star symbolizes elfin faerie kind), so anything having to do with humanity overall comes under the influence of this spirit. She acts to spiritualize matter, which is to say to awaken its inherent magic and essential possibility. She reminds individuals of their spiritual nature, and

167

of their destiny as spirits. If you need to awaken the spirit in others, she can be of great help in this regard.

While it is not entirely certain, the name Bismuth may come from the German words meaning *white mass*, since the name for elf *alb* comes to German from the Greek word for white, there is some connection there. Mass, of course, would mean a group, as well as, if we wish to extend it, a magical ceremony, and thus we may go so far as to say she is related to the Faerie Host (the white/silver mass) and the evocation of magic from the Moon and Stars and its marriage to the Earth magic that occurs during the Dance of the Faerie Circle. Thus the manifestation of Elfin on Earth. Something to consider.

Also, while she is a very stable being, this elemental is known to be weakly radioactive. This gives her a family relationship with the Fire/Radiance elementals, sort of a distant cousin so she can be used as a liaison to those spirits. While it is wise to call her within the magic triangle, she is not particularly dangerous to you (unlike Lead who is an out and out killer) and, in fact, if you are afraid and have a queasy stomach her presence will serve to soothe you. And unlike most spirits that tend to withdraw and become slow or reticent under cold conditions, she will actually become more expansive. You could call her in a walk-in freezer and she'd still be able to respond readily (check out the character Ty, the incarnation of the Norse god Hod, the God of all things cold and dark, on the New Zealand series the *Almighty Johnsons*).

Ancient miners called Bismuth *silver being made,* thinking it was transitioning toward becoming Silver. Somewhat in the same fashion that Alchemists thought they could force Lead to become Gold, only in this case the transition was occurring naturally. Or, perhaps more accurately, in the fashion that coal becomes diamonds.

This spirit had a relationship with the ancient Inca civilization. These days she is commonly found in Australia, Bolivia, and China and can be attracted by the magic of any of these cultures or used to influence these peoples. She has a natural affinity to rhombohedral lattice crystals such as dolomite and these can be used in her summoning. She is a big advocate of recycling and sustainability.

෴෴෴

The road to Elfin cannot be found on any map,
Nor is there a path that leads directly there.
It's winding, twisting passage is found more often
Through serendipity and whimsy
Than unbending intent and will power.

If you try to understand the elves, you are likely to wind up with them knowing everything about you.

One should always embrace an opportunity to learn.
—Old Elven Saying

The world is a potluck of magic.
Be sure to bring your own.

The true Nature of the Universe is infinite possibility.
You are just one of its many fabulous ideas.

Intolerance is un-elfin, except of intolerance for the intolerant.

Mere words do not wash one's actions clear

Elves frequently ask "why?"
But they also ask "why not?"!

Being an elf isn't a matter of belief, it's a result of Being.
We could be a leaf all day long
but we'll eventually fall from the tree,
but if we are the tree itself,
we'll live eternally through the flowering of the seeds.

The more you know about Life the more it surprises you.
—Elven Koan

CHAPTER FOUR:

THE LORDS AND LADIES OF FIRE, MASTERS OF THE REALMS OF RADIANCE, TRANSFORMATION AND ILLUMINATION

As we stated in the introduction, there are four states of being in Nature: solid, liquid, gas and plasma. While there are elements that can attain the state of being plasma, there are no elements that we know of that are plasma at normal room temperature and pressure. Therefore, we decided that radioactivity is a better corollary for the traditional elemental of Fire, since it is a state in transition and it is radiating. However, if you wish to use plasma as fire, simply do a little research and discover which elementals can achieve that state of being.

These are all dangerous elementals, think of fire, and we suggest they be evoked within the magic triangle with you in the circle of protection. As a general rule in dealing with these elementals, there are several forms of protection that may be of aid. Usually, one puts salt around one's circle, however, while that works well with most of the elementals of Gas/Air, Liquid/Water and Solid/Earth this is much less effective for these Radioactive Fire Spirits, unless it is iodized salt, which is harder to find these days, however, iodine (found in tablets, medicinal liquid and kelp) helps remove radiation from the thyroid gland (however, it is poisonous if overdone) and therefore is effective with these spirits. You can put it in any form around your circle or the triangle. Place the tablets around, dab the medicine on the borders or have it in glass bottles, even have kelp placed around your circle, or call the elemental of Iodine to aid in your conjuration.

One may also put glass bottles, or votive candles in glass holders around the triangle of manifestation and/or put bits of lead around your magic circle. Pencils won't do so well, because the lead in pencils is really graphite (we wouldn't want little children sticking Lead in their mouths) although, graphite does have some power to moderate these spirits. Since lead is often used as weights for divers, you can check a boat store if you can't locate lead elsewhere.

Technetium, Lord Arficor
(pronounced air – fie - core)
Lord of the Giants
Chant: "Great and wealthy you may be, but I'm your
equal you will see."

On the Earth, only minute amounts of this element can be found. However, man has been artificially creating this element as a by-product in nuclear reactors. In fact, the name Technetium comes from the Greek word meaning artificial. Obviously, although we don't necessarily recommend it, you could evoke this elemental near a nuclear reactor. However, remember it is not the element you are summoning but the elemental, the spirit of the element; it's probably best, when dealing with these very dangerous spirits, to summon them on your own territory, not their home ground. Keep your home field advantage.

This elemental is another being whose existence is prophesized long before he was discovered. He will appear wearing silver gray raiment and like all of these fire elementals he will have a radiating aura. He is the elemental of Giants, particularly Red Giants, which is to say aggressive, martial giants. Thus he has province over large corporations,

172

particularly those that are involved in hostile takeovers, nations that invade others for no valid reason, and individuals who use *power over* to attain their will. If ever you are in conflict or oppressed by bullies, individual or collective, this elemental will assist you. If you find yours'elf fighting against those who are far larger and greater than you, that is you have no choice but to take on superior forces and you find yours'elf totally out of your league, this is the elemental to call as long as you, in fact, are in the right and are being unfairly dealt with and oppressed.

This spirit also has some medical diagnostic abilities and can help you discover difficulties, or energies out of balance on the physical, mental, psychological, soulful, astral and spiritual planes of being. In that way, he is an ally to shamans. Also, if you are subject to psychic attacks on the astral plane but don't know the sender, he can trace the energy back to its source. He is also noted for having a keen interest in the functioning of the brain, myocardium, thyroid, lungs, liver, gallbladder, kidneys, skeleton, and blood.

Lord Arficor has an ancient relationship with Poland, especially that portion that was once known as Prussia, and with Sicily. He also has a relationship with California and the Congo from when it was known as the Belgian Congo, thus by extension with Belgium. These days he is known to reside in Ontario, Canada, and in Petten, Netherlands. All these peoples, areas and their related mythologies and magical traditions can be used to appeal to him. Hexagonal crystals can also be utilized to attract him, such as apatite, vanadinite, quartz and beryl among others.

He is a very magnetic being in many ways and has a wide, deep and penetrating field of influence. Bullies who take you on when he is your ally are really asking for it.

ৡৡৡ

**Promethium, Lord Pornacor
(pronounced pour - nay - core)**
Lord of Daring
Chant: "On the edge I'll ever be, alive, aware and
ever free."

There are certain individuals who love to live on the edge, to be in constant danger, to ever be a part of the *action*, to be at the cutting edge of life and society. Most of us prefer stability and only an occasional encounter with the cliff that plunges to depths unknown, but some individuals thrive on this atmosphere. They are often referred to as adrenaline junkies and they are only happy when in the midst of danger or excitement. If you find yours'elf in such a situation, have to deal with such individuals, or are, or wish to become one, this is the spirit for you. If you have ever met an individual of this sort you know that they can be exciting but also dangerous and sometimes amoral. Promethium is, of course, named after Prometheus the Greek God who dared to steal fire from the Gods and give its secrets to Mankind.

Daring, or *to Dare*, is one of the four vows of the Wicca. Thus this spirit has relationship to witches and magic users, especially those who dare to go beyond tradition and explore into new realms of magic. This extends to scientists and other explorers as well. Thus this elemental has some similarity to Ununoctium's Lord Murdorkere who is the Lord of the Unknown.

This elemental remained hidden for sometime, although there were alchemists who had seen his shadow and was certain he existed. He had been spotted in Italy and here and there in America, but he was finally tracked down in Tennessee. He is one of the lanthanide posse and surely the most dangerous of that crew, and by his nature constantly on the move. He has a natural relationship with Europium and Uranium. But, like those who do love to live on the edge, he is a very rare guy.

174

Like the other lanthanide's, he likes to wear silver white clothes, however, he is so fast and elusive you will probably only notice a greenish aura. It will be like looking at a green luminous ghost.

Both body centered and hexagonal crystals have an appeal for him and can be utilized in his conjuration. Hexagonal crystals include apatite, vanadinite, quartz and beryl and some of the body-centered crystals are pyrite, sphalerite and galena. He has an association with the constellation of Andromeda and the Greek Mythological figure for whom that constellation was named, who was, like Prometheus, chained to a rock. Her, to be devoured all at once by a sea monster (however, Perseus saved her) and him, eaten slowly and eternally. So any correspondences with Andromeda as a constellation or Mythological figure can be used in his summoning.

Polonium, Lady Bimocor
(pronounced bye - mo - core)
Lady of Harmony
Chant: "Release the tension, let it go, let friendship
        and compassion flow."

Have you ever met someone that you've haven't encountered previously who just seemed to hate you instantly or you felt the most incredible tension around? Sometimes that can lead to love, but often such static in relationship is an impediment to what one wishes to accomplish. The same can be true of groups that are brought together to achieve some goal but the personal energies and conflicts get in the way of the task. This elemental has the power to remove the static from such situations so that interaction is easier, smoother and more is accomplished. As we said, however, some folks thrive on this tension, consider it the spice of relationship, so be certain you wish to use this spirit's aid before you conjure her.

Appearing wearing silvery gray attire, she has the power of levitation and will probably hover a few feet above the triangle. You may see a bluish aura around her; however, if the atmosphere is hot her aura may seem pink or yellow. And she very well may be smoking a pipe, cigar or cigarette. She has a family relationship with the elemental of Uranium who might appear with her. Lady Bimocor also has dominion over urchins and gamine street children. All latchkey kids come under her protection. All those littles who are streetwise and pixie wild fall in her domain. The Baker St. Irregulars of the Sherlock Holmes stories would be hers as well. Her motto is *live fast and die young*. We suspect she likes the *Fast and Furious* movies.

This spirit has an affinity to France, particularly loves Paris, and to Poland for which Polonium is named. At the time this elemental was discovered, Poland, which had once been a great empire, was divided and under the control of Russian, German, and Austro-Hungary. The Curies who discovered Polonium hoped that naming it thus would aid in gaining Poland freedom and recognition and this elemental has made this so. She also has a love of Manhattan and the nightlife there (notice the Sherlock Holmes show *Elementary* takes place in New York City). Like most night people, she has an affinity to the Moon, so Moon correspondences can be used in her evocation. And like most wild children she has a hunger to explore and poke her nose into everything.

Remember enchantment is the key to elven magic. Don't try to pressure this elemental, she will only spit at you if you do. However, she has a love of tobacco so this can be used to draw her to the triangle. However, don't smoke this tobacco afterwards but crumble it and leave it in places you wish her influence to take effect. She is also known to have a love of seafood, if you wish to present that as an offering.

≈≈≈

> **Astatine, Lady Unsato**
> **(pronounced you-n – say - toe)**
> Lady of the Void
> Chant: "The Great Abyss I would to leap, to other worlds and magics deep."

*Y*ou won't see this elemental manifest in any physical form. She is for all intents and purposes invisible. However, you may notice that the lights dim and come up again when she arrives. You may also notice, if you look out of the corner of your eye, that the area above the magical triangle will darken as though it is absorbing all the light in the immediate area. All we can really say about this spirit's appearance, is that she is darker than black. If you can think of black somehow shining, that would be her. She comes from another realm and her manifestation in the material world is necessarily brief. The name Astatine comes from a Greek word meaning unstable, which refers not so much to her personality but her inability to manifest for other than brief periods.

This is the elemental of the Void, the Great Abyss. If you believe you are ready to pass the Rings Pass Not (Knot), then this is the elemental to evoke. Ninjas and their arts fall in her dominion, as do modern Goths and dark pixie folk. In fact, all individuals who dare to be thems'elves despite their society's disapproval come under her influence and protection. If you would know anything of these folks or seek to have influence with them, Lady Unsato can be of great help. She can teach you how to utterly disappear from the world if need be.

It is not certain, but some folks believe she gives birth to Iodine's Lady Rodfaso. There is certainly a family relationship. She also helps bring Bismuth, Polonium and Radon into the world. She further has friendly relationships with Sodium, Palladium, Silver, Thallium, and Lead and close friendships with the elementals of Boron, Carbon, and Nitrogen. She also

177

has an active association with Oxygen. The minerals hemimorphite, olivine (peridot), aragonite and marcasite can be used in her summoning. Cold water placed in the triangle has some power to help her to remain for a time.

This elemental has a love of Berkeley, California (a great place). It is possible this spirit had some influence on the anti-war demonstrations that took place in Berkeley in the 60's since this elemental urges us to leap the gap to higher realms of being and to set aside outdated traditions and thought-forms.

Francium, Lady Untov (pronounced you-n – towv)
Lady of the Seventh Chakra
Chant: "I would to integrate my life with all
of being without strife."

In some ways, this elemental's manifestation is the opposite of Astatine's Lady Unsato. That spirit draws light from your magic triangle; this one is like a sudden brief pulse of light, as though someone quickly reflected light through a mirror into your eyes. While the lights will tend to darken when Lady Unsato appears, Lady Untov will cause the lights to suddenly brighten momentarily. If you are using candles they may flare up when she arrives. The area within the magical triangle will seem brighter. However, Lady Untov will not stay for long. About 22 minutes is the length of her tolerance of the gross material plane, so welcome her, thank her and accept her gift.

For a gift is what she brings. She rules the seventh chakra that integrates the other chakra energies. If you call her and she responds she will flash your being, giving you a charge of energy that will set your body, soul, mind and spirit to integrating as a whole. She has no other purpose and while you can ask her for whatever you wish, this is what you will receive.

Her manifestation, however, will make your triangle more conducive to the conjuration of the elementals of Astatine, Radium, and Radon. They will be more comfortable responding once they sense that she has been there. And while most will only see light when she comes, those with psychic sight have said that she bears a resemblance to the elemental of Caesium, the time lord. Others have said she is like a circular rainbow with a pink aura at the center surrounded by red, then orange, yellow, green, then blue. She is known to be a descendent of Neptunium and Actinium.

This spirit has an eternal love of France both ancient and modern and the French people, their magic, mythologies and cultures. She is known to associate with the elementals of Uranium and Thorium. She is at ease with water and so a bowl or chalice of water may be put in the magic triangle for her. Pour it out in areas where you wish a blessing to come after your evocation. She is known to have a keen interest in alchemy and its descendant chemistry. So if you are an alchemist or chemist you will be doubly blessed by her manifestation. She functions mostly on the atomic level of being and is noted for helping to order the underlying nature of material reality.

Remember, her only purpose is to supercharge your energetic nature so your entire being will fuse together as a whole promoting your spiritual evolutionary development. It is likely that things will happen rather rapidly for a while after that, although some have been known to be simply stunned for a day or two. You may experience very vivid dreams for a time after her visitation. Pay close attention to these. You will surely begin to function on higher levels of magical being afterwards. The amount of friction you experience depends upon how readily you accept this new energy or your tendency to fall back into old habits. Don't summon her if you are not ready to embrace change.

❧❧❧

> ## Radium, Lord Alucor (pronounced a – loo - core)
> Lord of Cosmic Rays
> Chant: "My magic power greater will reach beyond
> the furthest hill."

This elemental will gradually appear into manifestation. At first, he will seem to be merely a luminous shimmering presence, then he will coalesce into a form wearing silvery white that will gradually darken the longer he remains until his clothes appear to be black with a yellowish tinge. Also, there will probably be no discernable aura at first but then slowly a dark scarlet/carmine radiance will become noticeable. When that happens, when he becomes black and carmine, it will probably be about time to release him. He is a very energetic and active spirit and is a bit restless. You don't want to hold him too long. Also, he often comes wearing a pocket watch; if he starts checking this frequently, you will know it is time to let him depart.

Lord Alucor is the Master of the Seven Rays of manifestation and, in fact, has province over rays in all forms, so even the idea of ray guns come under his influence. The seven rays are a rainbow of energies that are blended together to form the various aspects of intellectual, soulful, and spiritual development. If you wish to know anything concerning these rays or to use these rays in your magic, Lord Alucor is the expert on this subject.

The elementals of Uranium and Thorium are his frequent companions. The elementals of Barium, Strontium, and Lead are his colleagues. They work together to manifest various energies into the material world. However, he tends to start very simply, using one or two rays to initiate the process of creation. He regularly advises magicians to begin with the elements, the simple principles of magic, as a basis for the foundation of all that they would do. This elemental is the parent of the elemental of the gas Radon. He is also known to associate with the elemental of Beryllium.

He loves body centered cubic crystals and minerals like galena, pyrite and sphalerite, so these can be used in his evocation. And apparently, he collects watches, clocks and other chronometers. If you have a unique or antique watch, a cuckoo clock or some other unusual timepiece you can put that into your magic triangle to draw his interest. He also has an ancient relationship with France and is another elemental that loves Paris, a city that was founded by a Celtic people called the Parisii, so this spirit also has an association with the Celts. All these can be used to attract his presence. These days he is commonly seen in Belgium, Canada, the Czech Republic, Slovakia, the United Kingdom, and Russia and many of the counties that were previously Soviet states.

There have been some that believed he has curative powers, but this has been discredited for the most part. He has little interest in healing human beings; his main interest is in radiating energy and blending it to shape and illuminate the world, although, he does have a talent for detecting flaws. If you think you have some flaw in your magic or your personality and want to know what it is and how to overcome it, he can be of great help in this regard.

ॐॐॐ

Actinium, Lord Metacor
(pronounced me - tay - core)
Lord of the Blue Ray
Chant: "Genius wakens in my mind, I understand
all that I find."

This is the first non-primordial radioactive elemental discovered, which means that in comparison to some of the others, Lord Metacor is quite young, relatively speaking. Like the young, he takes a fresh look at the world and the Universe and tries to understand it anew. He is the Lord of the third ray of manifestation, the blue ray and manifests in the form of consciousness and intellect, thus he would be associated with the symbology of Mercury, and the sephiroth Hod on the Tree of Life of the Kabbalah. The third ray is often called the Ray of Active Intelligence and is sometimes also associated with Saturn, Binah on the Tree of Life, as the organizing principle of the Divine Magic. These correspondences can all be used in his conjuration. If you wish to improve your intellect and understanding, increase your genius, and particularly translate thought and ideas into action, this is the elemental for you.

Lord Metacor is the head of the actinide gang of spirits. He manifests outfitted in silvery white clothes with a golden cast but is most noted for the eldritch blue light that shines around him. He glows in the dark. He brings light to the dark places of the world and our minds. However, he tends to quickly protect himself from Oxygen and his silvery white robes will become pure white shortly after he manifests. Also, he tends to be very comfortable in the magic triangle and may settle in for a while. He loves to expound on his ideas and has a wide and connected understanding of life, the Universe and nearly everything else.

He will almost always be found with the Uranium and Thorium elementals. He is the grandchild of Thorium. What's a gang leader without his chief accomplices and supporters? Some people notice a similarity between him and Lanthanum's Lady Naena although she has a greenish aura while his is blue. They are both the offspring of the elemental of Uranium and are strongly attached to this spirit, and each the leader of their own gang. While she heads the lanthanide gang and he the actinides, his crew tends to be more diverse, liberal and open-minded. He has a love of France and Germany and their peoples, magical traditions and mythologies. He is also known to be fond of Sydney, Australia. Maybe he likes the opera? You may try playing opera music for his evocation. The name Actinium comes from the ancient Greek word for ray or beam, for this elemental beams his thought forms into manifestation, in this way influencing the world with inspiration and new ideas. If you wish to be inspired or inspiring, he can be of great help.

Like many of the younger spirits, he's very much into Science Fiction and promotes actual space exploration and travel. He's also known to be very concerned with the quality of water on the Earth. He envisions a better future for all of us and endeavors to inspire us with the ideas that can create that future.

৯৵৯৵৯৵

Thorium, Lord Charcor (pronounced char - core)
Lord of the Ancestors
Chant: "You sacrificed so much for me and now
I greatly honor thee."

Named after Thor, the Norse god of thunder, Thorium is a primordial element, and it and its elemental Lord Charcor are ancient. Thorium is also a grandparent of many of the other radioactive elements. He is older than the Earth and likely to outlive the Universe as we know it. He helps heat the Earth from within. This spirit rules the world of the ancestors, the old, and the process of aging and in that way has association with the elves as ancient beings and the oldest of the Fae, the Shining Ones. If you wish to connect with your ancestors to honor them and receive their blessings, if you need to deal with those who are old, or you wish to age gracefully, this is the elemental for you.

He'll first appear in bright silvery robes but these will slowly tarnish toward gray and then black the longer he stays. Some say he has grown soft in his old age and it is true that

Thorium is only weakly radioactive, being old, but don't let that fool you. Lord Charcor is still a dangerous old warrior. He has simply mellowed a bit with age. And it is said that although he is old he is still potent and fertile, however, he is very thin and skeletal in appearance.

Lord Charcor is more comfortable with squares than triangles, so have a square within your triangle of manifestation. He is known to associate with many of the rare earth metals. He has a lot of experience and is actually quite adaptable. He is noted for allying himself with the elementals of Chromium and Uranium. He is also said to hang with the Carbon and Phosphorus spirits. And although he is old he is still very active or some might say, as is often the case with the elderly, reactive. He is not that fond of fresh water, however, which tends to irritate him, so don't have water in your triangle. Although, he does have a fondness for the deep sea, so ocean water, particularly if it was gathered at sea rather than near the shore, can be used in his evocation.

As the name Thorium suggests, this elemental has a longstanding relationship with the Norse folk, the Swedish and the Vikings. He also has an ongoing association with the French, Polish and German peoples. Their culture, mythology, and magic are all very appealing to him. However, it should be noted the whole Earth is his bailiwick. But if he comes with attendants, which is possible and even likely, don't try to divide them so he is alone, he could become very fiery under this circumstance.

On the other hand, he has a reputation somewhat as a matchmaker. He has brought about more bonds between various elementals than any other spirit. He is noted for helping activate some very unusual relationships. Face centered and body centered cubic minerals can be used to summon him, such as pyrite, sphalerite and galena. Also wulfenite, rutile quartz and zircon can be used for the purpose. He is associated with Tarot major arcana #9 the Hermit or, as we elves call it, the Wizard, the ancient one carrying a lantern to light the way for others. He represents the wisdom of the ancients, the wisdom of age.

తిలిలిలి

**Protactinium, Lady Liurcor
(pronounced lie - your - core)**
Lady of Parents and Children
Chant: "Guide me as I guide you, loyal ever, ever true."

Protactinium means *first* or *before* because it is prior to Actinium and is the parent of that element; thus Lady Liurcor is the guardian of parents and children. Anything you wish to know about raising children, about parenthood, about the relationship between parents and children is within this elemental's area of expertise. Also, if you need to influence your children or parents or even the PTA, this spirit can be of great help. Like parents with children, Lady Liurcor is sometimes viewed as an unwelcome guest. But she knows the incredible magic of the childlike point of view, so if you're wise you will make her feel very welcome, even get very excited about her arrival. Make a big to-do about it.

Wearing bright silvery gray clothes, Lady Liurcor is a very adaptable being who interacts easily with the magic triangle, the square, and the five-pointed star. She has an ancient relationship with the Polish and Jewish peoples and anything having to do with their traditions, mythologies and magic can be utilized to appeal to her. By extension, she has province over Golems, homunculi, and robots, particularly as the latter develop Artificial Intelligence. She is known to have a deep interest in geology, in particular the development of sediments in the aging of the Earth and landscape. She further has an interest in the development and transformation of the oceans during the Ice Age as indicated by the various sediments. Several childlike spirits will probably attend her.

She is also known to have a fondness for Great Britain and its peoples and a love of Germany and its traditions. She has also been spotted hanging out in the Czech Republic and the Democratic Republic of the Congo. She interacts easily with water and steam so these can be placed in the triangle of manifestation. A pleasant smelling diffuser (it creates mist with scent) can be helpful. But she is even more comfortable in sandy soil, so sand, beach sand, loamy soil, clays, even your houseplant mix can be very comforting to her. In fact, the clays for a facial would make a great offering. The mineral wulfenite can also be used in her conjuration.

This is one of the rarest and most expensive of the elements; parenthood is not cheap; neither is caring for one's parents when they age. A true and happy bond between parents and children is, alas, also often rare, but this spirit can help you obtain that happy family state or certainly ease tensions and friction within the family or at family gatherings, thus promoting a more joyous atmosphere and improving relationships.

꒰ఠ꒱꒰ఠ꒱꒰ఠ꒱

Uranium, Lady Romicor
(pronounced row - my - core)
Lady of Strategy
Chant: "I set in motion forces vast whose wide effects
    will ever last."

Lady Romicor is a primordial elemental and while Uranium is weakly radioactive, it doesn't mean this spirit isn't dangerous. This is the spirit, after all, that destroyed Hiroshima. She was born in a supernova, so she is the product of tremendous power. She is in many ways a rich girl who has gone native. This elemental's great power, however, is in strategy. She is the Chess Master, the Go Master, the Shogi Master, the Mastermind, the great long-range thinker. If you wish to create a Domino Effect, setting things in motion so a series of events unfold, each causing the other until your magical will is achieved, she can tell you what domino to push with how much force, what magic to perform to set the whole series in motion (see the flawed but clever and very germane movie *Revolver* by Guy Richie, starring Jason Statham, also the great movie *Now You See Me*.). She is the master of the chain reaction. She has association with the second ray, which is most often called the yellow ray of Love-Wisdom.

This elemental is far older than the Earth but like many of her kin she will appear wearing silvery gray garments. She will most likely display a reddish-orange to lemon yellowish aura. She dresses in a sort of bohemian, hippie, gypsy style. She is the mother of many of the other elementals of the fire/illumination aspect and gives birth constantly. Thus her appearance will probably reveal that she is pregnant. By extension, she has province over expecting mothers. She is by nature both protective of those under her care and penetrating in the depth and breadth of her intellect and understanding. She is also one of the spirits that is fundamental to the care, birth, upkeep and life of the Earth. She is still in the process of shaping it according to her vision. And while she is clearly pregnant, most folks still think she is very *hot*.

Being named after the planet Uranus, correspondences of that astrological being can be used in her summoning. She has a love of the German and the French peoples and their magical traditions and mythologies, but also a love of the Italians and the Austrians. In fact, you could call her a very continental lady. She also has a love of America in its great diversity of cultures, and a love of Gabon in West Africa. In modern times, she is frequently seen in Kazakhstan, Canada, Australia, Niger, Namibia, Niger, Somalia, Uzbekistan, the United States, Argentina, Ukraine, China and Russia. Truth is, however, she likes to travel and has been seen here and there and nearly everywhere all over the Earth as well as cruising the oceans.

Lady Romicor is a somewhat sympathetic being, but not a pushover. She is unlikely to cave in to pressure and can be very hard if she needs to be, particularly if she is tired. If she appears exhausted, she will be very tough indeed. She interacts with nearly all the non-metal elementals, except for the noble gases whom she thinks are a bunch of snobs, which, let's face it, many of them are. Adamite, aragonite, heterosite, and tantalite are just a few of the minerals that can be used to attract her. Also, zircon, carletonite, rutile, scapolite, anatase, vesuvianite, narsarsukite, autunite, xenotime, thorite and zeunerite can be used as well as galena and pyrite. She likes crystals, what can we say? Also, earth in the form of sand or seawater can be used in the triangle to help stabilize her manifestation. But remember, like nearly all mothers, she can be dangerous if endangered. Good manners go a long way with her.

Neptunium, Lord Lolyni
(pronounced low - lynn - nigh)
Lord of Elven Star
Chant: "Ancient of days, ancient of ways, show me
the path through Life's murky haze."

Named after the planet Neptune, which was named after the Roman god of the seas, this elemental will manifest wearing metallic silver clothes. Naturally, correspondences for Neptune can be used in his conjuration. The sea he is the lord of, however, is not so much the oceans of Earth but the vast sea of starlight that fills the Universe. Lord Lolyni is the lord of the Elven Star, the seven-pointed acute angled star that is so beloved of modern Elven-Faerie folk. He is attached to the ancients, the Elder Race; not the ancient civilizations of Earth but the ancient civilizations and spirits that existed before the Earth came into existence.

188

Anything you wish to know of these primordial folk you can ask of this ancient elemental. Unlike the seas of Earth, the oceans of starlight are associated with fire and this element and elemental is indeed pyrophoric, tending to ignite spontaneously in the air. In fact, his silvery robes will most likely seem as though they are on fire. You may say he has a fiery aura. It is also said that he has very prominent cheekbones. He is both hard and adaptable, both we expect due to his long experience. The truly strong tend to be the most adaptable in these elves' experience.

This spirit has been reborn many times. His original primordial form has long passed and he's reincarnated again and again. If you wish to know about the process of reincarnation, of which he knows a great deal and has much experience, he would be happy to inform you. He and nearly all those very early spirits have moved on to new and, in most cases, more advanced spiritual lives.

This spirit has a relationship with Berkeley, California (where these elves read tarot on the streets a time or two). He also has a fondness for what was once called the Belgian Congo and therefore by extension an association with Belgium, which was founded by the Belgae, who were a mixture of Celtic and Germanic peoples. He has ongoing relationships with both the elementals of Uranium and Plutonium. It is said that, like Sherlock Holmes, he is a bit of an amateur detective. He is particularly adept at spotting the exceptional that has hidden itself among the mundane. If you need to know who are the secret wizards, sorcerers and magicians in your area, he can spot them easily. While some think that he can be used as a weapon, he has, so far, simply refused to be used in that way. He is too old and wise to involve himself in the petty and provincial feuds of the Earth folk.

Like the elemental of Uranium, this spirit loves crystals and celestine, chrysoberyl, danburite, and marcasite are just some of the minerals that can be used in his conjuration, as well as, cassiterite, carletonite, pyrolusite, and also galena and pyrite.

189

Plutonium, Lady Radasit
(pronounced ray – day - site)
Lady of Avalon
Chant: "A paradise in distance lays, where I will
      spend eternal days."

Originally named Hesperium, a word from the Greek meaning *the land in the west*, Lady Radasit is the elemental of Avalon, Erasea or Paradise, the promise of a magical world that lays to the west. It is the promise of a better day, better land, better world for all of us. If you aspire to create or live in a more perfect world, if you wish to participate in a Utopian society, then this is the spirit that can help you achieve that goal. However, this is also the elemental that is chiefly responsible for destroying Nagasaki, so she is incredibly powerful and not beyond dispensing with the old in order to make way for the new (see Tarot major arcana #16 the Tower). Keep that in mind when you evoke her. It may be that things you hold dear, cherished ideas you cling to will have to be let go before the vision you desire can unfold.

Faerie, some feel, is a land that exists parallel to this one and to which, like Narnia, one might escape; although it is still a land with dangers and trials for the spirit and the soul. Elfin is a land, like Rivendell and Lothlorien, made manifest upon the Earth. This spirit has power to aid one in both and either of these endeavors. If you wish to step into a parallel world or create paradise on Earth, this elemental is the one for you. There will, however, most likely be fallout from your magic. Such world transformational magics evoke chain reactions of events that affect all our lives. Once such spells are set in motion they become a powerful force and it is hard to adjust them, so know what you want before you initiate these energies. The commonly used adage *be careful what you wish for* is germane here.

Her robes will appear silvery white at first but quickly turn to dark gray as she manifests. It will seem as though she exudes a powdery dust that ignites in the air as she moves. The motion of her hands, for instance, will leave trials of fire. You may possibly glimpse her yellowish green aura. Those who have seen her say she is a bit of a femme fatale. She was first brought to this world in secret just as the path to Avalon was and still is occult knowledge. The way to the *promised land* is revealed by the prophets, visionaries and sages. At one time, it was thought that Plutonium would be the last element and thus the final elemental to be discovered, rather like the idea that paradise is the last and most perfect world, but that was not so. So, too, paradise/Elfin/Faerie demands continuing effort toward improvement. There is no end except our never-ending journey toward perfection.

This is another spirit that has a fondness for Berkeley, California, and for Tennessee. She also has a love of Rome and its beautiful old houses and winding streets. The Russians have courted her for years (read *Declare* by Tim Powers). The element is named for the planet Pluto so correspondences for that astrological body are appropriate for this elemental's summoning. Any of the crystals mentioned for the elementals of Uranium or Neptunium can be used for this elemental as well.

Americium, Lord Zårduron
(pronounced czar – dure [as in endure] - roan)
Lord of the Mists
Chant: "I seek the portal to the land where I find
my loving band."

The old saying *where there's smoke there's fire* is perfect for this spirit who is great at detecting smoke or in esoteric terms the *mists of Avalon*. If you are looking for the wardrobe that leads to Narnia, the fountains that let one into Fillory (see *the Magicians* trilogy by Lev Grossman), or the portal to some other realm or parallel world Lord Zårduron can be of great help. He can see the thresholds, the bridges and other passageways that enable one to enter other realms. He has the capacity to detect that faint whiff of magic, that subtle shimmer in the air that indicates that the world is just slightly different in a particular area and that one may step from one dimension to another if one is able to see the stairway to heaven at the right moment. He is not the only spirit that can help you move from one plane to another, but he is the very best at spotting the secret doorways.

Americium is named after the Americas, North and South, which were once known as the New World, and this spirit can introduce you to new and amazing worlds of magic and wonder. Although, he will also tell you that you may need to step from one to another to another to get where you ultimately desire to be. To get to the big city (the Emerald City of Oz?) you may have to go through a lot of small towns.

His clothes will be, like most of his metallic kin, a silvery white. You may observe a dark, red brown aura around him, rather like he is an ember quietly still burning inside. He appears to be quite soft and gentle in his way, with a quiet disposition and actually tends, which you realize when to get to know him, to be rather self-critical. He's in many ways harder on himself than anyone else. He's much less inclined toward

192

the explosive, bombastic style of the Plutonium and Uranium elementals. However, he does share with them a love of Berkeley, California and its very liberal atmosphere. He is also fond of Chicago. He likes the wind coming off the lake. And like many of them, he likes the Manhattan nightlife but usually visits there in secret.

However, he is, of course, one of the actinide gang, but is perhaps the most reticent of that crew and while he is less well known he does take part in their gang activity. So don't think that because he tends to be quiet for the most part that he's not dangerous.

Americium was originally called *delirium* because this elemental's reluctance to be separated from his gang was driving the investigating alchemists mad. Like many introverted, shy beings, he prefers to stay in the background of those to whom he feels a close affinity. At the same time, we might point out that stepping into other worlds, where the customs and principles of being are so very different from those to which one is accustomed can in itself be a bit maddening at times. The rules and passage of time in Faerie, Narnia, Fillory or Middle Earth are not always the same as here.

He does have a fondness for all crystals, so these can be used to summon him, but this is true of most of the actinide gang. He is not really fond of life on Earth, he'd much rather be elsewhere, so simply tell him of your desire to visit this or that world and release him. He will let you know when you get close to the portal or inspire your movement toward it, guiding you through inspiration and synchronicities. He does, however, like to swim in rivers and streams, so you may have river water in the circle for him so he stays a little while, but again, it is best if he doesn't stay too long.

Curium, Lady Ancor (pronounced ane - core)
Lady of Daimons
Chant: "My demons I master, my world is much vaster."

Initially, this element was called pandemonium, a word that was cobbled together by John Milton for use in *Paradise Lost*. Pandæmonium comes from Ancient Greek word for pan or all (and the Great God Pan and his son, Peter) and the Late Latin daemonium *evil spirit* or *demon*, from Ancient Greek *daimon* (inspiring spirit) or *deity*, which is also related to the idea of genies or djinn. Lady Ancor is knowledgeable and familiar with the djinn and daimons that can aid and inspire us and she can help you obtain a helpful daimon to speak to you of brilliant ideas in whatever area you desire whether you are an artist or scientist or both (we recommend Jonathan Stroud's *Bartemeus* books, also the brilliant psychology book by James Hillman *the Soul's Code*).

This elemental will manifest wearing silvery clothes with a bright, glowing purple aura. Like many of her posse, she favors Berkeley, California as a particularly inspiring place and does much to promote the various artists and scientists, also witches, there. She is further known to visit Chicago sometimes. And like the elemental of Americium, she loves Manhattan, especially the secret and out of the way places of which only the cognoscenti are aware.

She is, of course, another occult elemental whose manifestation was first held in secret. She is a hard and dangerous spirit, so just as a reminder, as with all these fire/illumination spirits, the magic triangle is advised. However, she is relatively calm as a spirit and unlikely to lose her temper easily. She can, however, get things going with very little input of energy and can sustain a magic spell for a long time with a small amount of investment. Whatever you offer her, if accepted, she will use to its upmost. As a magical investment, you get a lot of *bang for your buck* from this elemental. She also is said to hearten those who have begun to give up hope, whose hearts/faith/spirits are failing so to speak and give them new life and a steadfastness to endure. She is further acknowledged to have a scientific interest in the exploration of our solar system and our neighboring planets. And like her fellow gang members, she has a love of crystals of many kinds, so use the ones that will indicate to her the areas where you desire inspiration.

While Lady Ancor is an attractive being, she is also an enchantress who alters her response according to her reception. If she is received warmly, she can be very sympathetic, but cooler receptions, or a very reserved summoning, where one is clearly uncertain or distrusting toward her, will find her responding in kind. We're not saying you should simply trust her unreservedly, rather we suggest that you should do your best in your conjuration to be positive, confident and courteous. Be friendly and welcoming while keeping your wards in place. In this sense, she is different than Americium's Lord Zårduron who is naturally sympathetic no matter how you seem to be. She can be much more resistant than he is. However, she is another spirit that has reincarnated a number of times since her primordial beginnings and we expect she is still evolving, so she understands your aspirations to improve yours'elf and your life.

Berkelium, Lady Lonliyon
(pronounced lone – lie - yone)
Lady of Those in the Know
Chant: "I would to find those wiser ones, to guide me
till my work is done."

Rather like some of the denizens of Berkeley, for which this element is named, Lady Lonliyon doesn't seem to do anything in particular, or have a purpose other than to hang out with her friends. However, one would be mistaken to think her useless, for this elemental knows nearly everyone who knows anything and everything, so if you are in need of a spirit to tell you what elemental knows what and what abilities they have that aren't revealed in this tome, then she is surely the one to conjure. She knows all their secrets for she is a great listener and nearly everyone feels comfortable opening their hearts and minds to her. If she feels you can be trusted, she will surely reveal a thing or two to you. People love to confide in her, often without realizing they are doing so.

Besides being born (reborn) in Berkeley and spending a good deal of time there, she has also spent time in Tennessee and even Russia now and again. She loves the vodka and the caviar (in Tennessee she drinks Moonshine and in Berkeley we are told she likes a toke or two). She even spends some time in Idaho, what she does there we don't know (tips cattle, maybe?). She loves to hang out with scientists (she's a nerd groupie) and listen to their theories and the tales of their research. She is also the mother, though few but scientists know this, of the elemental of Californium.

She has a soft appearance and like most of her kind prefers silvery white garments. She has a deep red, nearly infrared, fluorescent aura. She *is* a bit of a groupie, and does tend to conform easily to group pressure, however, this is also why they tend to take her for granted and reveal so much to her without realizing they are doing so. Thus, she can teach you

196

how to fit in, how to be the *fly on the wall*, how to be an overlooked helper who hears everything because no one notices you. She is the patron of the Roman Emperor Claudius. If you need to go unnoticed (and thus safe) while you develop yours'elf and your abilities, she can be of great assistance. Some have thought to use her as an energy source, as they have used her cohorts, but she's a bit too laid back for that. Theoretically she could do it, but she's just not up for it. Maybe another day.

Like nearly all of the actinide gang, she adores crystals of various sorts so you can have whatever kind pleases you in the magic triangle, she is sure to enjoy it. You can also use rock salt in her triangle. And like others of her kin, she has been reborn many times since the Earth came into existence, having long abandoned her primordial state. She is regularly reinventing herself to fit in wherever she wishes to be. She can teach you to do the same.

ॐ ॐ ॐ ॐ

Californium, Lord Fluercor
(pronounced flew – ear - core)
Lord of Taoism
Chant: "With Nature I would reunite and this way set
the world aright."

If you are sick of the modern world, in particular the hustle and bustle of city life, then this is the elemental for you who would gladly help you re-integrate with Nature, with your own nature and Nature at large. If you would like to abandon the modern world for a simpler life, get back to Nature, live in the country or the woods, Lord Fluercor would gladly assist you. He is the patron of Buddha as he sits under the Bodhi tree seeking enlightenment, of hermits in their withdrawal from the world, and of wizards in as much as they live in Nature and involve themselves in their studies, rather than the affairs of the world. Radagast the Brown from the *Hobbit* would fall into this category. Lao Tzu illuminates this basic philosophy.

Curiously, this spirit likes to reside in Berkeley, California, a place that integrates both city life and Nature. So you don't have to retreat totally from the world to receive this elemental's assistance, you simply need the burning desire to reconnect with Nature, despite the influence of the modern world and to resist its attempts to turn you away from your genuine nature into an artificial being. If you wish or need aid to help you become your true s'elf, this is the spirit that can best serve you. It is true that doing such will make you stand out in the world, that you will become an example of individuality for others, and you will mostly likely be scorned by those who do not dare to deviate from society's norms, but such has ever been the way of the fae folk and surely will ever be. This spirit will also show you, if need be, how to appear to conform without losing contact with your true nature, how to join in without selling out (see Alice Hoffman's book *Practical Magic* or see the movie).

In manifestation, this spirit will appear wearing silvery white robes. You should be able to see him clearly, as though he were another human standing in your triangle. He will look strong, muscular without being muscle bound; his is lean muscle. Despite that, he is a rather yielding fellow and really rather adaptable as a spirit, altering his attitude according to the circumstances he encounters. There's a certain ease about him that makes normal folk uncomfortable yet he's known to draw the heavier elements to him, all of whom are rather awkward, artificial beings, but who most likely harbor an inward desire to be more natural.

He will be at ease in the magic triangle, but you could also put a square within it and that would also be okay. He has a love of homemade magic wands, if you have one or more of these you can put them in or around the triangle to attract his interest (see *the Witch's Guide to Wands* by Gypsey Elaine Teague). He tends to be more interactive when it is warm. So keep your magic room warm but not hot or cold, unless you just want him to come and chill for a while or get so excited that he rants on and on about his favorite topic, which is getting back to Nature.

Like many of his kin, he's also spent a lot of time in Tennessee, also in Savannah, Georgia and some time in Idaho (who knows what they do there, perhaps just hang out in Nature and get away from it all). Additionally, he has been seen in Russia now and again. Remember Russia is named after the Rus, a Viking folk, and thus associations with those peoples and their traditions and cultures appeal to him. And like most of the others of the actinide gang, he's fond of a variety of crystals.

He is, of course, a wizard, which means dangerous, as are all of the fire/illumination elementals. He is not fond of water, but he does like soil, so you can have some soil mix in the triangle that you can spread around later in areas where you wish his magic to take effect. He has the great ability to get things going, so if you are feeling down or hesitant or procrastinating about beginning something, he can stir the energy and help set things in motion with very little investment from you. He is a very efficient spirit and can keep a spell in motion for a long time.

Einsteinium, Lady Damducor
(pronounced dame - due - core)

Lady of Imagination

Chant: "I imagine a better world and in so doing it is unfurled."

---

    *E*instein, after whom this element is named, is often quoted as saying, "If you want your children to be intelligent, read them fairy tales. If you want them

199

to be more intelligent, read them more fairy tales." Many think that Einstein was a great mathematician, but in fact, most of his insights came from his imagination as he contemplated the nature of the Universe. This elemental has the power to increase one's imagination, to make it better, stronger, greater, wider, more detailed and alive. If you wish to improve and increase your imaginal abilities, this is the elemental for you.

This is an artificial element, in as much as scientists forced its birth, unknowingly, through the testing of Atomic Bombs. We've seen that in fact through the course of time/space the simplest elements have given birth to more complex elements and this continues to be the case but also these elements in transition also tend to age and transform into simpler elements, their essence is the same but their formulation unique. So, too, our imaginations give rise to worlds that did not exist prior to our imagining them. We give them birth, so to speak, are co-creators in the Universe. Just as Tolkien created Middle Earth, so do we create Faerie worlds and other realms and give them birth by living them. Lady Damducor would happily help you to bring your own realm to life in the world.

She is, it goes without saying, a dangerous spirit. Imagination can be dangerous. Dreams can be dangerous; there are nightmares (just ask Freddie) as well as healing dreams. Faerie isn't called the Perilous Realm for nothing. And magic has long been noted for its perilous nature so keep that in mind when evoking her powers.

Similar to her gang mates, she will manifest wearing silver raiment, only hers glows. She has a soft aspect to her personality, will be very comfortable in the magic triangle, and is noted for carrying a crystal wand. She is very empathetic and is associated with the High Priestess, major arcana #2 of the tarot, all correspondences for that card can be used in her summoning. Despite her soft appearance, she is a high energy being who can sustain a spell for a long time. Some think that she is merely a dreamer and has no practical use as an elemental, but that would be like saying Einstein was useless because he spent most of his time dreaming and theorizing.

This elemental has a strong love of the Pacific Islands and in that way has association with the Menehune folk of those lands. Like the rest of the gang, she is known to hang out in Berkeley, California. But she also has a love of Sweden and its peoples, cultures and mythologies. Pyrite and galena can be used to attract her as well. As she ages, she takes on the aspects of the elementals of Berkelium and then Californium, so she is strongly related to, admires and is supportive of them. It is wise with this spirit and really with all the fire/illumination elementals, to fast before conjuring them, and especially not to eat or drink during their evocation. They haven't come to feast but to set things in motion. These are serious beings.

෴෴෴

Fermium, Lord Famyn (pronounced fae - men)
Lord of Awakening
Chant: "I've come to sense I'm something more than how I thought mys'elf before."

Forced into this world in the modern age by a hydrogen bomb, which is in a sense a very short-term manmade sun, this elemental signals the awakening of the new aeon. Like some of the others, he has long since abandoned his primordial form, reincarnating again and again. While Pagans and Christians say they are reborn (which is strange since Christians don't believe in reincarnation), we elfin faerie folk speak of awakening the dormant elfin magic within. This elemental is a Master at helping individuals to awaken. If you need help in your own awakening, awakening being a process not merely a momentary explosion of consciousness, or if you wish to help others to awaken to their true natures, and the stars know there are many who desperately need awakening, Lord Famyn will surely serve you well.

This spirit will be content in the magic triangle and like the elementals just previous will be wearing or carrying a magician's wand. He has a fondness for the Tropical Islands in the Pacific, where he was reborn, but is also sometimes seen in Berkeley, California. He further has a love of Sweden and its peoples and mythologies. He is additionally known to spend time in Tennessee and Nevada. He feels especially comfortable in Gabon in Central Africa, where he feels like he can truly be his natural self, and any association to the original peoples of these areas can be used to attract him. Like some of the others in the gang, his existence was shared at first among alchemists and sorcerers who kept his life secret in order to utilize his power for their selves. He is associated with the number 19 and all correspondences for that number, the number of the Sun in the major arcana of the tarot, can be used in his conjuration.

In appearance, he will seem to be a dwarf, gnome, hobbit or even more likely a menehune or some other small elfin being. His taste in clothes varies and unlike so many of the others in his gang, he will not necessarily wear silver or silver-white or gray. Rather, he is more inclined to wear camouflage of a sort (of course it was elves who first invented camouflage in order to hide in the forest, adapting it from our observations of Nature). Have you ever seen one of those photos where a person has been painted to appear like the background behind them, or one of the aliens in the Predator movies? This is what he will seem to be: there, but hard to distinguish from the area around him.

Of course, being a somewhat new elemental, which is to say, new to this world and our knowledge of the Universe, there is much we don't know about this being, whose nature is evolving and whose power evokes change and awakening by its very nature. We are just getting to know each other and are barely beyond the introductory stages of relationship. You may discover many new and interesting things by evoking this being, both about him and yours'elf.

क़•क़•क़

> ## Mendelevium, Lord Tosoto
> ## (pronounced toe – so - toe)
> Lord of Alchemy
> Chant: "I will to change the world itself to create a land of love and health."

This and the elementals that follow may be hard to see when they manifest. Their appearance often will be more in the form of a hologram or virtual reality simulation that is somewhat hazy or wavering. In a sense, Lord Tosoto's manifestation is like a holographic television signal that is beamed into the magic triangle. He's not really there, but you are seeing his image on a magical Skype. This is due in part to the fact that these elementals are newcomers to this plane of being and are just beginning to find their way here as well as fashion the form with which they will choose to manifest. Like those immediately before him, he will carry a magical wand, only this one will be a glass stirring-rod for the various chemicals he combines. He is known to like vodka and other hard liquors mixed with water, so these can be placed in the magic triangle to draw him.

Lord Tosoto is the Lord of Alchemy and thus has association with all alchemists and chemists and the art of alchemy whose essence is the attempt to transform one material state to another, which is also to say to help elementals transition to new positions of power and expertise. These spirits, just like the Shining Ones and like ours'elves, are evolving and at times undergo such profound transformations as to seem to be utterly new, as differentiated from their previous selves. This is not to say, however, that in

203

transforming they will not become like some other, higher, more evolved being who has, most likely, moved on to more advance planes of manifestation and spiritual development hirs'elf.

This element is named after the great Russian alchemist and diviner Mendeleev, who created the Periodic Table of Elements in order to understand and predict the nature of the elements and the elementals. Therefore, this elemental has a strong association with Russia and in particular its magical and scientific community (you might enjoy the wonderfully magical *Night Watch* series by Russian writer Sergei Lukyanenko). Like the other members of the actinide crew, Lord Tosoto is also fond of Berkeley, California. And since the word Alchemy, comes from the words Al – Kham, Al being a form of El, a signifier of the Shining Ones and enlightened being and Kham was a region approximately where Tibet is, only larger, all associations to that region and its cultures and magic can be utilized in his conjuration. The idea of Shambhala or Shangri-La as hidden paradises on Earth and as the Ealds of the Ascended Masters or Shining Ones would be appropriate correspondences here, also in that sense would be Avalon.

Like Plutonium's Lady Radasit, he is associated with the number 16, the number of the Tower in major arcana of the Tarot, since he represent attempts to force transformation. However, unlike her, he isn't trying to destroy the old in order to make way for the new, but to transform the previous condition into the newer one. It is the difference between revolution and working to change the system from within.

In most cases, it is best to let things develop naturally. But there are times when one can simply not stand by and let things go on as they have been. Magic, it is true, is a function of Nature and works within Nature's laws and is the application of will and intent to create change and indicates force or applied effort by its very nature within these natural laws. However, here a good deal of force or pressure is to be used and this elemental can tell you how much can be utilized without destroying the Tower, which would create chaos rather than improved conditions. This improved condition is the true goal of Alchemy. Of this and much more this spirit can inform you. As the old saying goes, *you don't want to throw the baby out with the bathwater.*

కావావా

Nobelium, Lady Lonicor
(pronounced low – nigh - core)
Lady of Peace
Chant: "I set aside my sword most sharp to play
upon the magic harp."

This element is named after Alfred Nobel, the inventor of dynamite, who then went on to establish the Nobel Prize for excellence in Science, Literature and achievements in furtherance of humanity. Thus this elemental is the master of transforming the arts and sciences of war and destruction into the creation of civilizations that promote peace and harmony. Correspondences to Mars can be used to conjure this elemental, but not in his aspect as the God of War but rather in his more ancient aspect as the God of Farming. This elemental can teach you how to turn swords into plowshares, turn hostility into friendship and move your individual world and the world at large toward more peaceful enterprises and activity.

There are those who will tell you that war and conflict are part of mankind's nature, but this elemental will inform you that this idea is merely an assumption not a fact of Nature. This notion is as invalid as the idea that dogs and cats are natural enemies and can never get along. She will help you find the ways to a more peace-filled and successful life and how to use peace as an active power and magic to achieve your goals. How at the center of magic, like the center of a hurricane, there is a peaceful calm place and if you can create this center of peace and harmony within yours'elf the world will begin to swirl around you fulfilling all your desires.

The elemental has a sororal relationship with the elemental of Ytterbium, the elemental of Precision. She will not appear, however, if you don't have an aqueous solution, such as water, in your magic triangle; she needs this to formulate herself. She will carry a wand of power, entwined with ivy, but she will appear to be a being made of liquid. She is not naturally opposed to the magic triangle; however, it is hard for her to manifest without this liquid. You may find her standing on one leg, as in a yoga pose or she may have her palms toward you as a symbol of peace and goodwill. In this sense, she looks somewhat like a Hindu goddess. She, in fact, has associations with all goddesses and gods of peace, including Tara, Kuan Yin, Nienna, Tolkien's elven goddess of mercy, and the Roman goddess of peace and tranquility, Tranquillitas, and images and correspondences of these beings can all be used in her conjuration.

She has an association with Sweden, Russia and the United States and their many peoples, traditions and cultures. She has spent years endeavoring to bring peace and reduce conflict and competitiveness between Russia and the U.S. In her summoning, pick those god-forms that are most indicative of the area to which you wish to bring peace. Like others of her club, she is also quite fond of Berkeley, California.

Lawrencium, Lady Mynodcor
(pronounced men - node - core)
Lady of Acceleration
Chant: "More quickly I do now evolve and
every problem I will solve."

Like Ununoctium's Lord Murdorkere, Lady Mynodcor will be moving very quickly when she appears. She is the Master/Mastress of acceleration and tends to get things going faster in your life. The idea of the Quickening from the *Highlander* series of books, television shows and movies, is similar to the effect she produces. She activates your spiritual being. She initiates you into a new level of operating. You begin to function on a higher speed, frequency and vibrational level. Your elven faery nature comes to the fore and asserts its'elf and you begin to evolve very quickly. This may be overwhelming at first, think of individuals when they are first becoming vampires or werewolves, etc., or even high school students first acclimating to college, but as you get used to this new state and way of being, you will find that you are not only more powerful but have a great deal more freedom.

While it is true that she moves or vibrates at high speed, you will probably see her as a swirling spiral of energy, rather like a silvery colored dust devil twirling in the center of your magical triangle of manifestation. She has an association with the number 12, the number of the Hanged Man in the major arcana of the tarot, so correspondences to that card can be used in her evocation. However, this also means that no matter how much you wish the quickening to occur she won't initiate it unless you are truly ready. This is in your best interests after all, for otherwise you could suffer terrible consequences. At the same time, don't worry, if nothing happens at first, the conjuration is not wasted for she will surely leave you with an energy burst that will ignite when the time is right. You will be like a sleeper agent waiting for the password that will set you in motion. In this case, it will be the circumstances that signal that your spirit is ready for this enormous influx of power and energy.

This elemental has a family relationship with Lutetium's Lord Ariscor, although they hang out in different crowds. While many expect her to be, because of her power, quite volatile, as we said, this is not necessarily so. If the time is not right, she will not take any action, except preparatory action.

Like many of the others immediately before her, she is fond of both Russia and the United States, and in particular Berkeley, California. She is, however, the final member of the actinide gang and the other elementals who follow will be of another sort. She is fond of crystals, like most of the rest of the gang, in her case hexagonal crystals are a particular favorite. And she is also at ease with water and other liquids and these can be used to help her manifest. If you put water in the magic triangle she may appear as a waterspout that one might encounter at sea.

ॐॐॐ

Rutherfordium, Lord Nelbarcor
(pronounced kneel- bare - core)
Lord of Spiritual Transmutation
Chant: "To my core I do transform and thus
in all ways I'm reborn."

This elemental has association with the numbers seven and four, seven being the number of the elven star and a prime number and four the number of solidification and manifestation. This is also the first spirit of a new group of elementals, although this group is not so much a gang as similar beings who happen to hang out together, more of a social circle or coterie so to speak. This is the elemental of spiritual transmutation, which is to say the alteration of the individual spirit at its very core. This is not mere physical change, or transubstantiation, although it often involves that; it is deeper than shape-shifting in which one changes one's shape but remains essentially the same; but is rather the change of the heart, mind and soul of the spiritual being.

Laws can be made to curb the hate and prejudice of certain individuals in a society, but such laws cannot change the tendency for prejudice and hateful behavior in the first place. If such societies are to change, the individuals themselves must change, first their behavior but ultimately their prejudices, beliefs and opinions that incline them toward such behavior. This elemental can help you alter your own inner nature to a more spiritually evolved state of being, but also help you to promote such transformation in others. He is an alchemist, of course, but an alchemist of spiritual alchemy, the transformation of the Lead of our beings into the Golden light of realization. He is not the only elemental with this power; however, when it comes to a deep, enduring and thorough change, he is the very best at this.

In the Christian faith, the wine and bread of communion are mystically transubstantiated into the blood and body of their Christ (their most evolved member of their faith and the ideal toward which, if they are truly Christian, they strive). While this may seem like symbolic cannibalism, it is not much different than the idea that one can drink the blood of a vampire and become a vampire. Except that this is a spiritual transubstantiation. The wine and bread isn't physically changed into the Christ's blood and body, but are mystically transformed (unlike the physicality of the vampire blood). You are consuming the essence, the spirit of the Christ in order to become a better (in their case, more Christian) being. It is an elevation of spirit and that is what this elemental does. He helps transform your spiritual energy so you may function on higher, which is to say more subtle, planes of life, becoming not merely an elf or faerie but a Shining One.

Lord Nelbarcor has a natural association and family relationship with the elemental of Hafnium, the Lord of Dragons. He deals with the great and powerful and can teach you how to do the same, although again his energy is more spiritual than worldly. It is rather like the Dalai Lama visiting Presidents and Premiers. He is with them but not entirely of them. This elemental also has a natural affinity to Russia and the U. S., particularly the city of Berkeley in California. He also has a love of hexagonal crystals and you can use these in your magical triangle.

If you burn incense, and any incense that evokes spiritual feelings in you can be used (Nag Champa is recommended, but frankincense works for many folks but whatever works for you), he will use this as a medium for manifestation. However, what you will probably see will depend upon your own spiritual inclinations. Hindus would most likely see a Hindu god-form, Buddhists Buddha, Pagans one of many various gods or goddesses depending on their disposition, and elves perhaps one of the Shining Ones.

๛๛๛

Dubnium, Lady Ruyntyn (pronounced rue – in - tin)
Lady of Quantum Leaps
Chant: "I cross the great abyss it's true and find that
I am ever new."

Like Lord Nelbarcor, Lady Ruyntyn's appearance will depend upon your natural inclinations and expectations. How you think she will be will greatly affect how she seems to be. She is the elemental of Quantum Leaps and can help you take a significant step forward to a new level of operating, whether you are a scientist, artist, magician, musician, tennis player, auto mechanic, computer geek or whatever. She can reveal secrets and techniques to you that will up your game and improve your ability in any field that you desire, although she does have a particular love of alchemists, scientists in general and atomic physicists.

She has a natural and family relationship with the elemental of Tantalum, Lord Harala, the Lord of the Cosmic Carrot, which is to say the elemental who urges individuals to evolve with a promise of great rewards. She has an especial love of Russia and her traditions, peoples, mythologies and magic, but like many of these later fire/radiance elementals, also has a love of Berkeley in California. She additionally has a fondness for Denmark and Germany and their cultures.

The numbers seven and five are sacred to her, so the combination of the seven-pointed elven star and the five pointed wiccan star can be used in her evocation. Thus you can have a five-pointed star or seven-pointed star within the magical circle and triangle of manifestation and she will find this very appealing. We suggest that if you are of elven faery kind you have an elven star in your magic circle and a five-pointed star in your triangle, but if you are wiccan but not of the fae folk, have the reverse.

This elemental is known to promote the development of fusion over fission as an energy source and seeks to help scientists make the leap to this improved technology. However, her manifestation on the Earth is still very limited as, consequently, is her power to influence those who manifest here.

Seaborgium, Lord Åtovåcor
(pronounced ah – toe – av [as in avocado] - core)
Lord of Shamanism
Chant: "Nature holds the secrets true to all we will
and wish to do."

This is the elemental of Shamanism in its manifestation as the precursor to Alchemy, Chemistry, and particularly Medicine. This spirit knows everything about herbal remedies, Bach's Flower Remedies, Ayurvedic medicine, Chinese herbal medicine and even Pharmaceuticals as they find their roots in Nature. If you wish to know about the power of plants and even of these elementals to be combined for healing or enlightenment (spiritual and psychological healing), then this is the spirit to conjure for he can tell you all you wish to know on this subject. By extension, this elemental is also the Master of healthy eating, thus preventative healing practices and can help determine a diet that will best serve your body and your needs. He emphasizes the necessity for living foods, which is to say foods that provide vitality to the individual and haven't lost their potency.

Lord Åtovåcor will seem to be a very quick spirit. His appearance will be quite brief, no more than a minute or so, however, he will most likely seem to be a weathered old man with a broad smile, who will manifest, listen carefully to your request (be succinct) and perhaps nod or otherwise signal his understanding and then depart. He knows what he is doing and you will surely hear from him, in terms of receiving the results of your questions or requests, in due time. This response may come from inspiration, dreams, or some circumstance or individual who enters your life. His form when he appears may be overlaid with illuminated geometric patterns. You may briefly feel like you are having a psychedelic vision or looking through a kaleidoscope upon viewing his form (see the visionary art of Alex Grey).

The spirit has a natural affinity to Tungsten's Lady Bramyl, who is the Lady of Sorcery, whose emphasis is more on psychology than the physical body, but these are clearly related, the shaman also being a psychological healer, although he often uses herbs and plant spirits to evoke this healing (see Terence McKenna's book *Food of the Gods: The Search for the Original Tree of Knowledge – A Radical History of Plants, Drugs, and Human Evolution*). He has an association with the numbers seven and six and thus with the unicursal hexagram, the six-pointed star as it is drawn in a continuous line as some magical frasorities (fraternity/sorority) formulate it. He has a love of Russia, particularly Moscow, and Berkeley in California but he also has been known to favor college towns wherever they are found.

Like Dubnium's Lady Ruyntyn, this elemental promotes the development of fusion as an energy source for the future. And, of course, he urges you to eat well for the benefit of your body and your spirit.

&&&&

Bohrium, Lady Danicor
(pronounced day - nigh - core)
Lady of Destiny
Chant: "Beyond this life worlds await and destiny has
set the date."

Some people confuse Fate and Destiny, and while they are intertwined they are not the same. Fate has to do with the circumstances you face in life, both those that are karmically ascribed to you but also all things that happen to occur on your path. Destiny, however, is the evolutionary direction you are headed in. It is the ideal s'elf that you are bringing to life in the world and no matter what fate you may encounter, which can delay your destiny, your destiny will eventually unfold. Fate in that way is the road you travel

213

toward destiny, which is your destination (you can see how those two words destiny and destination are linked).

Lady Danicor can tell you all you need to know and all that you are prepared for and ready to understand about your Destiny (There will be things that cannot be revealed until you have developed certain higher realizations. She could tell you but you just wouldn't understand.). She will inform you of what you need to know to undertake the next part of your evolutionary journey, however.

This spirit is associated with the number seven in a double sense. Correspondences to major arcana #14 Art, in the tarot can be used in her conjuration as can #17 the Star and card #7 the Chariot. She aids the awakening of the true s'elf, which is to say one's magical and spiritual s'elf in every individual whomever they may be. Like the previous elemental, she will not manifest for much more than a minute. She will appear to be much like the Star card, in that she will seem to be the shining form of a woman holding two cups, one held upright and the other downward. However, this initial image will quickly begin to fade until she disappears altogether. Have what you want to know written out, or even better created on an image/vision board or collage and she will get the message and will do her best to align your desires with your destiny and vice versa. Do not be afraid of your desires, you are who you are because of who you are; yet your destiny may require you to change.

Like the others of this circle, she has a fondness for Russia and Berkeley, California, but she also has a great love of Denmark and its peoples, cultures and mythological magics. She has a sororal relationship with Rhenium's Lady Radencor who is known as the Lady of the Skies. All correspondences to heavenly goddesses can also be used in her evocation, such as Anat, Isis, Innana, Astarte, Hera, the Nordic Frigg, especially, and even Mary of Christian mythology if you happen to be a Christian elf, witch or magician, which does occur, although not often (check out the Hallmark series *the Good Witch*). She also has a love of hexagonal crystals so these can be put in the magic triangle.

ॐॐॐ

Hassium, Lord Majicor
(pronounced may – ji (rhymes with high) - core)
Lord of Star Mates
Chant: "Together we will rise toward light, as our spirits do take flight."

This spirit has a natural relationship with Osmium's Lady Luwyn who is the Lady of Soul Mates. The elemental, however, rules Star Mates, which are sometimes but not always one's Soul Mates. Certainly, one's soul mates are nearly always also star mates, but while soul mates tend to indicate a romantic and feeling relationship, a blending of hearts, star mates indicate those individuals who are destined to help one evolve as a spirit and that one in turn helps to further on their path toward s'elf realization. It is mutual evolutionary assistance, but also Dharma or responsibility. Star Mates often have a mission to fulfill together. One's Star Mates nudge one toward one's destiny and an encounter with a star mate is sure to set magic unfolding. If you wish to know anything concerning your star mates, or draw them closer to you or you to them, this elemental can assist you in that regard.

Lord Majicor has a natural affinity to Germany, particularly the area called Hesse and its history, culture and magical traditions. He also has an association with the Russian people and their traditions. The numbers seven and eight are sacred to him, so any correspondences to these numbers, including the eight pointed star-like symbol of chaos magick can be used in his conjuration. The tarot card, Strength, major arcana #8 and all its correspondences can also be used to attract him. While his appearance will be brief, he does have a silvery star-like aura. One gets a sense of a very powerful, yet eccentric, being. And he will also appear to have some deformity, most likely a slightly atrophied leg, which will make him stand slightly askew. He has association with Richard the Third, the deformed king of England and with Hephaestus the Greek god of blacksmiths, craftsmen, artisans, sculptors, metallurgy, fire and volcanoes. Hephaestus' Roman equivalent is Vulcan and his Elven form from Tolkien would be Aule, the god of the forge. And by extension from all these god-forms, this elemental bears a correspondence to dwarves as dark elves, particularly in their ability to forge magical artifacts. In Norse mythology, he would be Brokkr, the dwarf god, but he also has association to Wayland the Smith of Teutonic myth.

He is a highly magical being, with an affinity for magic numbers (so magic squares can be used in his conjuration) who reinforces one's sense of uniqueness, since one's Destiny is one's own and in a sense unique to each individual. He will tell you how your unique nature, even your foibles can be a source of magical power. This spirit is magic to his core. He is also very strong and unlikely to yield to pressure of any kind. It is said of him that he is as strong as a diamond; think of a blacksmith. Be courteous always but not obsequious. Request his aid politely and with dignity and he will certainly help you. He is further associated with the platinum group metals so he is both wealthy and evolved and surely has a sense of his own self-worth.

Meitnerium, Lady Nuficor
(pronounced new- fie - core)
Lady of Star Enchanters
Chant: "The magic of the stars we weave and
what we will we shall achieve."

L ady Nuficor is the patron of Star Enchanters, those Shining Ones who weave the magic of the stars for the purposes of enchantment. This is a particularly elven magic. Her sacred numbers are seven and nine, the nine-pointed star being the star of the star enchanters. And all associations with these numbers can be used in her conjuration. In Chinese mythology, nine is linked to the creative power of the Dragons. The Dragons in this case being the Star Spirits. Anything you desire to learn about Star Enchantment, this spirit knows and if asked politely would be glad to share with you.

This elemental has a relationship with Germany and Austria and their cultures and mythologies. She also has strong ties to the elementals of Cobalt, Rhodium, and Iridium. Her manifestation will be brief but she will seem to be a flickering star being, that is to say, she will have the outline of a person that is filled with stars. She is a noble being of high stature and evolutionary development. However, like the elemental of Hassium, she is a very unique, eccentric and individual being. She's not a Barbie doll girl. Her beauty comes from the uniqueness of her being and her vast intelligence and genius. Most normal folk would find her strange if they could meet her in the flesh. They would be very spooked by her presence and yet intrigued without knowing why.

She has a fondness for face centered crystals, such as pyrite and galena and these can be used in her evocation. Like her neighbor the elemental of Hassium, she is strong and unlikely to yield to pressure. Treat her like a dignitary come to visit from an important foreign land and you will find her to be a very

217

sympathetic being, after all she is an enchantress, and quite receptive to your inquiries. She is also a very magical being and she loves the magic of numbers and math and its power to understand the Universe.

჻჻჻

Darmstradtium, Lord Wihòcor
(pronounced why - how - core)
Lord of Luck
Chant: "Lucky I shall ever be, fortunate to be so free."

This is the elemental of luck. Not the elemental of lucky accidents, like Palladium's Lady Hasarcor, but of luck in general, what we elves like to call elf luck (As an example of elf luck, yesterday we went to send a package to a friend in appreciation for some kindness she had done. The package cost us six dollars to post, but right after that as we sat on a bench down the hall from the postal store, organizing ours'elves before leaving, we looked down and saw what looked like stickers under our bench, this turned out to be $10 worth of stamps.).

You could also call it Cosmic Luck, since it depends not upon serendipity so much as the development of the individual spirit and soulful nature. The more attuned you become to your true nature and thus to Nature overall the luckier you become. You could also call this luck, magic, not the magic of will power, except in as much as one has willed and made effort to become a better person, but the magic of synchronicities, the magic described by the bummer sticker when it says, "Magic Happens".

Lord Wihòcor is associated with the numbers seven and ten. Ten is the number of the Wheel of Fortune in the major arcana of the tarot, and all correspondences to that card in its positive aspects and to the number ten as a number of

218

completion is relevant here. He has a natural relationship with the elemental of Platinum, the elemental of the High Elven, those most lucky and fortunate of elves. But he is also known to have a relationship with the elementals of Nickel and Palladium. Consider these things when you evoke him.

He has a love of Germany in particular and its cultures, peoples and mythologies and die hexerei (witchcraft), die Zauberei (sorcery, magic, wizardry, witchcraft and conjuring) and magie (magic) but also a love of science in general, as he views luck and magic more as a science than an art. Like many of his social circle, he has a thing for magic numbers, the Code of Life, and magic squares can be placed in the triangle of manifestation to attract his interest. Use one that relates to the area where you would like luck to manifest for you.

This elemental will not appear for very long, rather like luck, but will come and go in a matter of minutes or even seconds and you may not notice much more than his noble bearing. If you do see him, he will probably have a huge smile on his face; after all, he is an extremely lucky fellow. However, even though his presence is brief you can be sure that he has brought you a bit of luck that will manifest in due course. The most important thing really is to keep developing yours'elf and your magic, for that bit of luck is not like a piece of cake that you eat and then it is gone, but more like a piece of fruit that you can enjoy and be nourished by but which also bears seeds that you can plant for the future. He has a love of octahedron shaped crystals, such as diamonds (we have an amethyst in this shape), so these can be used in his conjuration but also body centered crystals and minerals can be effective.

Roentgenium, Lord Oloråcor
(pronounced oh – lore - rah - core)
Lord of Insight
Chant: "To the depths I will perceive and
none shall ever me deceive."

This elemental holds the numbers seven and eleven to be magical and sacred. The word for eleven in German is elf, which makes this power particularly elven in nature. Eleven is the number of Justice in the major arcana of the tarot and all correspondences to that card and number can be utilized for his evocation. Naturally, to achieve Justice, to be a Judge, requires great insight if one is to do it well, but this spirit doesn't merely inspire Justice, although that is surely part of what he does, but he also fosters insight overall. Whatever field you would like insight into, whatever person or situation you'd like to understand, or if you'd just like to increase your insightful nature, this elemental can surely assist you.

Lord Oloråcor has a familial relationship with Gold's Lord Urm who is the Lord of the Common Weal. The Common Weal requires great and profound insight. It requires a visionary ability that few have as yet developed. He also has strong links to the elementals of Copper and Silver. You can learn quite a bit about a spirit by the company sHe (she/he) keeps.

This spirit spends a good deal of time in Germany, which he loves. It is, in fact, his favorite place on Earth, and like most of the others of his social circle, he's not really very comfortable on Earth as yet. Alchemists keep summoning him, trying to create the conditions that are pleasing and conducive to him, but he, so far, still demonstrates a reserve that is often the case of those of noble birth. His manifestation is very brief and yet he will inspire you with insight into whatever category you wish. Perhaps, you will use it to understand him better.

Surely you can use it to better comprehend any or all of these elemental beings.

What you may see in that brief period of manifestation is a flash of silver. What will stand out, however, will be his eyes that will seem to bore into your very soul. In the fleeting moments that he appears he will quickly come to understand you, your desires, destiny and evolutionary needs and his gaze will penetrate to the depths of your being. Face-centered and body-centered crystals can be used in his evocation, as he has a fondness for both of these. Unlike those just previous, however, he has no particular interest in magic squares. You could use these in his evocation if you wish, but that would be like trying to attract a fanatical football fan with baseball memorabilia. Hex signs, such as you may find among the Pennsylvania Dutch, can be quite efficacious however.

Copernicium, Lady Télegålåcor
(pronounced tell – lee – gal - lah - core)
Lady of Astrology
Chant: "I would to understand the world,
           to which my spirit has been hurled."

This is the elemental of astrology and, in fact, its brother astronomy as well, although astronomy likes to pretend they are not related, acting as though astrology is crazy, wears a hat made of tin foil or is ever away with the faeries. In truth, however, astrology is merely another language, like mathematics, for attempting to understand the nature of the Universe and human beings' place within it, and attempting, by virtue of its observations and principles, to make some predictions about the future. That astrologers are often no more accurate about this than the weather forecasters is to be explained by Chaos Theory. There are simply so many

221

variables that it is hard to make predictions with such a limited fund of data. Yet, despite that fact, these elves have met some amazing astrologers who have quite accurately told us our sun signs even though they had never met us previously nor elicited any information from us before doing so. If you wish to know anything about astrology or astronomy, this is the elemental to summon. If you wish to understand others or a particular situation using astrology, she can also be very helpful.

As one might expect, this elemental has a relationship to the numbers seven and twelve, twelve being the number of houses in a western earth centered astrological chart and twelve being the number of sun signs, also the number of Chinese zodiacal signs. All correspondences to these numbers can be used in her conjuration, including major arcana #12 of the tarot, the Hanged Man that signifies our dependence upon the forces of the Universe that shape our lives and development.

This spirit is, alas, not entirely comfortable in the magic triangle. However, she is at ease in squares, so having a square within the triangle is a good idea in order to facilitate her manifestation. She has a natural association with the elementals of Zinc, Cadmium and Mercury so keep in mind that they have certain similar tendencies, however, she is a far heavier, more serious being than they, perhaps because they are older and less inclined to take themselves as seriously. She has been especially noted for similarity to Mercury's Lord Merku, the Lord of the Words of Magic, who is her cousin. You might try calling him or the elementals of Zinc or Cadmium prior to evoking her since being an extrovert she is extremely nervous and unstable if she is alone. She loves to be on the go and to have things going on around her. She is also likely to form close working bonds with the elementals of Copper, Palladium, Platinum, Silver, and Gold. Yet, while she is similar to these other elementals in some ways, she is still a unique and highly eccentric individual. Don't think that because they have similarities that they are exactly the same. They are not.

Like most of the others in her social group, she has a love of Germany and its ancient peoples, myths and magics but also a love of Japan and its peoples, magic and culture (see the two movies *Onmyoji* about Abe no Seimei the famous Japanese magician and astrologer). Naturally, being the Master of Astrology, she has a natural affinity to Cosmic Rays and their powers.

Water works well in the magic triangle/square for her evocation. This can be poured out after the evocation in areas where you expect or wish her powers to manifest. She is also a noble being, of ancient heritage, although still just beginning to display her powers on Earth. Despite the fact that astrology has been a part of human culture for aeons, it is still in its infancy as a science (however, we highly recommend Dane Rhudyar's book *An Astrological Mandala: The Cycle of Transformations and Its 360 Symbolic Phases* if you wish greater understanding of evolutionary astrology).

When manifesting, she will appear to be a metallic gaseous being, looking rather like an Amazon warrior woman made of mist. What? Did you think all astrologers were nerds? She is known to have a fondness for hexagonal structured crystals, so you can use these in her evocation.

ॐॐॐ

Ununtrium, Lady Murdorcor
(pronounced muir - door - core)
Lady of Polyamoury
Chant: "I embrace all those I love with passion wrought from stars above."

The name for this element is a temporary placeholder and will change in the future as soon as the scientists concerned can agree on one. The elemental itself only comes here briefly and only when evoked because for the

223

most part the Earth and the people here are not quite ready for her influence. This is the elemental of polyamoury, which is the power of loving and sustaining relationship beyond the pairbond without evoking jealousy. This is not an easy thing to accomplish as yet in most cultures on the Earth, and this elemental is thus highly unstable here. However, anything you wish to know or learn about sustaining multiple relationships you can ask of this spirit who is more than happy to teach you the principles upon which such relationships can be maintained, the most significant of which are mutual respect, loyalty and spiritual aspiration. She can help you establish an island of stability for such relationships in a world that, for the most part, is hostile to such "marriages". She can also tell you of polygamy: polygyny in which a man has multiple wives simultaneously; polyandry, where a woman has more than one husband at the same time; and group marriage in which the family unit is composed of multiple husbands and wives.

This elemental has a natural affinity to Boron's Lord Burnas, the Lord of Formation. She also has relationships with the elementals of Aluminum, Gallium, Indium, and Thallium. She has a love of Russia and her peoples and mythologies, as well as Berkeley, Ca. in the U.S. and a fondness for the Japanese people and their magical mythologies (we recommend the movie *Mushi-Shi* or the magna or anime series of the same name for a bit of Japanese magic).

Lady Murdorcor moves at near the speed of light, so her manifestation in your magical triangle will seem to be no more than a flash of starlight, which in a sense it is. Having some object that gives off steam in the magic triangle can be quite helpful in allowing her to stabilize temporarily. Long enough, at least, to state your desires and requests. She has an affinity to the number seven, a number sacred to the elves, and all correspondences to that number can be utilized in her conjuration. And she additionally has an association to the number thirteen, the number of Death and Transformation. Polyamorous relationships will surely transform the individuals involved but will also affect society overall and these relationships signal an advancement for humanity as it progresses toward greater spiritual evolution.

৵৵৵

Flerovium, Lady Soficor (pronounced so - fie - core)
Lady of Initiation
Chant: "The mysteries will be revealed and
Nature's secrets thou will yield."

Lady Soficor is the Master/Mastress of the Magic Theatre, the Master of Ceremonies of the Mysteries of Initiation and she would gladly aid you to develop your own initiatory powers, which is to say not only to initiate you into deeper mysteries, depending upon your current level of development, but also inspire the powers of initiation within you. You will become, if you are not already, what is often called a s'elf starter, an individual who gets things going based upon hir (his/her) own initiative. Her sacred numbers are seven and fourteen, and all correspondences for these numbers can be used in her evocation. Fourteen is the number of the card Art in the major arcana for the tarot, and this is because the mysteries are an aspect of Art and are often enacted and revealed through theatre. If you desire to hold initiation ceremonies, she can help you to make these truly potent and magical, rather than merely empty and meaningless gesture and ritual as so many ceremonies are. She will instill your ceremony with true initiatory power and the individual will be transformed through the experience of initiation.

This elemental is a noble being and will only appear in etheric form. Her presence will be like looking at a hologram. You will see her and see through her (see Salvador Dali's painting the Last Supper). She will seem to be essentially a luminous egg. She has a natural association with the elemental of Lead, the spirit that rules the military. Like that spirit, she is highly organized and can help you initiate a vortex, coven, lodge or other magical group. One can learn from her how to create their own eald (elven group and magical space), their own island of stability in an uncertain world. However, because her area is initiation into the mysteries, she also has association with mystics and their direct communion with the Divine Magic. She exudes magic. While she is organized, she doesn't support authoritatively hierarchal organizations but rather promotes the idea that each individual should be the authority in their own position of expertise. This is the elven way, initiated cooperative action. In that way, this spirit has doubly magic possibilities, she teaches one how to lead/initiate/guide and follow/cooperate simultaneously. Perhaps it takes an elven mind to understand this.

This elemental has a love of Russia and its peoples and mythologies (see Russian director Andrei Tarkovsky's film *Stalker*), but also has a fondness for the Germanic peoples and myths. These can be used to attract her. But it is important to understand that she is essentially a cosmic being and attachment to the Earth isn't really about any particular group or culture but about the development of advanced spirits and moving each one closer to greater realization of hir (his/her) true being.

Ununpentium, Lord Murdoråcor
(pronounced muir – door - rah - core)
Lord of Feng Shui
Chant: "Closer I am everyday to the land where I may stay."

This is, of course, only a temporary name for this element, but the elemental's name is Lord Murdoråcor, signifying his relationship to a number of other elementals in this group. The numbers seven and fifteen are sacred to him and all correspondences to those numbers can be used in his evocation. Fifteen is the number of the Devil card in the major arcana of the tarot and, indeed, this spirit helps one make a choice between good company and bad as well as distinguishing what places or areas are best, most fortuitous and most conducive to one's success and spiritual evolution and what regions would merely block one's energy, hold one back or even tempt one to regress.

This is the elemental of Feng Shui, the Chinese art and science of finding a place in the world where one can live in harmony with Nature. It is, in a sense, the art and science of location, location, location. (And in that sense, this elemental has province over Realtors.) Although, it additionally teaches one how to make the most of the situation one finds ones'elf in to heighten the possibilities, harmony and opportunities wherever one may find ones'elf. This elemental also has knowledge of Astro-cartography, the art and science of using astrology to determine those places on the Earth most fortuitous to the individual. All of this and more this spirit will eagerly share with you.

This elemental has natural family relations with the elementals of Nitrogen, Phosphorus, Arsenic, Antimony, and Bismuth, although he is a much heavier and serious, cosmic being than these spirits. He is much less Earth oriented than they, and his power, while it can be used to find your proper place upon the Earth, is really about you finding your true place in the Cosmic Order, although in many ways we are not talking about your proper place in the Universe but rather your true velocity, trajectory, frequency and vibration. Of this and more this spirit can inform you. He will teach you how to be mobile and stable at the same time. In that sense, he has an affinity to the Romany folk and when he appears, moving at near the speed of light, he may look to be a gypsy man streaming by.

He has a love of Russia and its peoples and is working very hard to stabilize their world and economy so it will be the true paradise they have long envisioned and sought to create. He also has a fondness for the Germanic cultures.

Livermorium, Lord Ralicor
(pronounced ray - lie - core)
Lord of Cosmic Civilizations
Chant: "Greater worlds I'll come to know, into the stars
I will now flow."

This elemental can inform you of everything you wish to know concerning Cosmic Civilizations, the civilizations among the stars: past, present and to come (read Frank Herbert's *Dune* series). Cosmic civilizations, however, refer not only to beings that have extended thems'elves beyond a particular solar system, but also those that have evolved into other dimensions and parallel worlds. Of all this and more this spirit can tell you (see Timothy Leary's *The Game of Life*).

Lord Ralicor has association with the numbers seven and sixteen. Sixteen is the number of the Tower in the major arcana of the tarot and represents the periodic rise and fall of civilizations. This elemental will reveal to you how a civilization may be maintained through constant evolution, improvement and development so that it will endure for aeons.

This elemental also has enduring familial relationships with the elementals of Oxygen, Sulfur, Selenium, Tellurium, and Polonium. He additionally has a fondness for California in the U.S. and Russia and its peoples but also a strong and long standing relationship to the peoples, cultures and traditions of Poland. All these countries have had alchemists and magicians who have evoked this spirit and drawn it to the Earth.

He is a very swift being, moving near to the speed of light, this due to his cosmic nature. And while he is new to the Earth, he is an ancient being and quite committed to his purpose of helping create great civilizations among the stars. Part of what he is doing on Earth is encouraging us to branch out, become stable enough and harmonious enough to extend into space, and as part of that he promotes the development of nuclear fusion as an energy source. He also urges the blending of peoples, cultures and races to create a greater, stronger and more harmonious society. In the future, he will encourage us to merge with other humanoids we encounter in space to create a more powerful hybrid species.

This elemental will probably be seen carrying a wizard's staff with a jewel at the top that radiates light. He will not look entirely human, although his appearance will be humanoid. He will have pointed ears and be dressed in what may appear to be sacerdotal attire. He is comfortable with water so a chalice of water can be put within the magic triangle that you can pour out later in areas where you wish his influence to manifest.

> ## Ununseptium, Lady Ėldåcor
> ## (pronounced el - dah - core)
> The Star Queen
> Chant: "Shed your grace on me, great one, and guide me
> to the worlds to come."

This elemental is known as the Star Queen. Her name, Ėldåcor, literally translates as elf-ium in Arvyndase (see our book *Arvyndase [SilverSpeech]: the magical language of the Silver Elves*). She'd be the equivalent of Varda in Tolkien's elven mythology. Her sacred numbers are seven and seventeen. Seventeen is the number of the Star in the major arcana of the tarot and like that card, this spirit brings blessings to all those who seek to create a better world. Her mere presence will bring a blessing upon you and your life and she will help you in any way possible to enlighten you, illuminate your being, make you greater, more powerful, successful, healthy and improve your life and your magic. She is, in her way, the most advanced and powerful of elementals and there is very little that she can't achieve for those to devote thems'elves to truly improving the world, making it more beautiful, magical and harmonious, while simultaneously endeavoring to reduce conflict, prejudice and factionalism in society.

This spirit will be around long after the current Universe has collapsed and begun to reformulate itself. She is very close to being immortal.

Lady Ėldåcor has a natural and familial relationship with the elementals of Fluorine, Chlorine, Bromine, Iodine, and Astatine. She has been evoked with some success by alchemists and magicians in Russia, Germany and the U.S. and has encouraged all these folks to work together toward the greater good and the benefit of the Earth and its inhabitants.

You are not likely to perceive this being in a visual way. However, if you feel very blessed, healed, encouraged, inspired

230

to create a better world and more confident in your abilities then you will know you have been successful in your evocation.

ઌઌઌ

Knowledge is achieved by the mind.
Wisdom is born from the heart and gut.
—Old Elven Knowledge

There are those among occultist who speak of Storming the Gates
of Heaven and taking them by force.
This is not true of Elfin.
One cannot force their way into Elfin any more than we can
compel someone to love us.
Elfin bestows its favours when it will and we can but make
ours'elves ready by becoming worthy and suitable lovers.

Elves don't keep pets,
We keep friends.

The Elfin language is not primarily a language of words but of
tones. One can utter elfish words and never be speaking Elven, yet one
could speak any language in the Universe with a sincere hear and be
speaking the secret language of he elves.

One of the greatest gifts of the elves
is our propensity for giving presents.

There was a time when men first came and razed our forest for
their farms that they would leave us milk and biscuits and other
offerings outside their doors at night to placate us and offer tribute.
Now they offer our elfin kin food stamps,
which is neither as personal nor poetic,
but at least the milk is fresh.

It is unwise and futile to attempt to deny
a fool their right to suffer.

A spiteful tongue strangles it's owner from within.

If I knew then what I know now,
I wouldn't be so ignorant at this moment.

CHAPTER 5:

IN CONCLUSION

Will There Be Other Elementals?

Of course. There are already predictions and speculations about elements number #119 and #120 and other greater/heavier elements and thus elementals to come. We know that these elementals, rather like us, breed and create new and, we hope, greater and more advanced versions of their selves. They produce sons and daughters and often this occurs in quite passionate encounters. We also know that they age and transform through time and experience. Surely, there are more to come and as we ours'elves advance in knowledge and experience, we will be able to conjure them more effectively.

Is It Necessary to Use the Magic Circle of Protection and the Triangle of Manifestation?

No. You are not calling the elements to you but the elementals. If you were dealing with the elements themselves, the magic circle would be the equivalent of and replaced by a Hazmat suit and the triangle of manifestation would be like a glove box where you could manipulate materials without being contaminated or the containment area in nuclear power plants where the fuel rods are safely (hopefully) stored.

However, you are not dealing with the elements in most evocations at all (although you could use Gold or Copper, for instance, in the conjuration of those elementals), but rather with the elemental spirits and thus the magic circle and triangle represent psychological and psychic barriers and boundaries. Some individuals are secure enough in thems'elves and advanced enough to perform such conjurations totally in their minds. On the other hand, there is so much in the popular culture as well as magical tradition about protecting ones'elf from demons and the dangers of failing to do so that it is easy to doubt and be uncertain and these can be very powerful and dangerous spirits in that sense. So unless you feel utterly secure in your own psyche, so that no doubt and second-guessing can enter in, the magic circle and triangle provide a very helpful psychic barrier. They are a form of insurance. They are not protecting you from the elementals so much as from your own shadow, your own demons manifesting as fear and doubt and the demonic introjects (read about Object Relations Psychology) of certain individuals that promote such uncertainty. However, ultimately, it's up to you. And you will always have our Elfin blessings for your safety and success.

ABOUT THE AUTHORS

The Silver Elves, Zardoa and Silver Flame, are a family of elves who have been living and sharing the Elven Way since 1975. They are the authors of 34 books on magic and enchantment, including: *The Book of Elven Runes: A Passage Into Faerie; The Magical Elven Love Letters, volume 1, 2, and 3; An Elfin Book of Spirits: Evoking the Beneficent Powers of Faerie; Caressed by an Elfin Breeze: The Poems of Zardoa Silverstar; Eldafaryn: True Tales of Magic from the Lives of the Silver Elves; Arvyndase (Silverspeech): A Short Course in the Magical Language of the Silver Elves; The Elven Book of Dreams: A Magical Oracle of Faerie; The Book of Elven Magick: The Philosophy and Enchantments of the Seelie Elves, Volume 1 & 2; What An Elf Would Do: A Magical Guide to the Manners and Etiquette of the Faerie Folk; The Elven Tree of Life Eternal: A Magical Quest for One's True S'Elf; Magic Talks: On Being a Correspondence Between the Silver Elves and the Elf Queen's Daughters; Sorcerers' Dialogues: A Further Correspondence Between the Silver Elves and the Founders of the Elf Queen's Daughters; Discourses on High Sorcery: More Correspondence Between the Silver Elves and the Founders of the Elf Queen's Daughters; Ruminations on Necromancy: Continuing Correspondence Between the Silver Elves and the Founders of the Elf Queen's Daughter; The Elven Way: The Magical Path of the Shining Ones; The Book of Elf Names: 5,600 Elven Names to Use for Magic, Game Playing, Inspiration, Naming One's Self and One's Child, and as Words in the Elven Language of the Silver Elves; Elven Silver: The Irreverent Faery Tales of Zardoa Silverstar; An Elven Book of Ryhmes: Book Two of the Magical Poems of Zardoa Silverstar; The Voice of Faerie: Making Any Tarot Deck Into an Elven Oracle; Liber Aelph:*

Words of Guidance from the Silver Elves to our Magical Children; The Shining Ones: The Elfin Spirits That Guide You According to Your Birth Date and the Evolutionary Lessons They Offer; Living the Personal Myth: Making the Magic of Faerie Real in One's Own Personal Life; Elf Magic Mail, Book 1 and 2; The Elves of Lyndarys: A Magical Tale of Modern Faerie Folk; The Elf Folk's Book of Cookery: Recipes For a Delighted Tongue, a Healthy Body and a Magical Life; and Faerie Unfolding: The Cosmic Expression of the Divine Magic.

The Elven Way is the spiritual Path of the Elves. While all elves are free to pursue whatever spiritual path they desire, or not as the case may be, these elves are magicians and follow no particular religious dogma. We understand the world as a magical or miraculous phenomena, and that all beings, by pursuing their own true path, will become whomever they truly desire to be. You can always contact us through our website at: http://silverelves.angelfire.com or join us through our Facebook page, under the name Michael J. Love or you can try The Silver Elves (Facebook has a way of changing our name from time to time).

www.ingramcontent.com/pod-product-compliance
Lightning Source LLC
Chambersburg PA
CBHW051451170526
45166CB00001B/198